Introducing Genetics

R. N. Jo[...]

A. Karp

Department of Biochemistry
[...]thamsted Experimental Sta[...]
Harpenden

John Murray

© R. N. Jones and A. Karp 1986

First Published 1986
by John Murray (Publishers) Ltd
50 Albemarle Street, London W1X 4BD

Typeset, printed and bound in Great Britain
by J. W. Arrowsmith Ltd

British Library Cataloguing in Publication Data

Jones, R. N.
  Introducing genetics.
  1. Genetics
  I. Title II. Karp A.
  575.1   QH430

ISBN 0-7195-4235-9

Preface

This is an attempt to write the kind of textbook that we were always looking for as sixth form pupils but were never able to find. It contains all of the elements of genetics that are needed for the 'A' level examinations in biology and related subjects, i.e. botany, zoology and human biology, up to the special paper level. In addition it should satisfy the requirements for various BTEC programmes as well as covering the basic material in many polytechnic and university foundation courses.

The text deals with all of the factual information required at the Advanced level. There are worked examples of genetic problems, and a wide selection of questions and problems from recent past papers. The book sets out to explain the concepts involved and gives clear and concise definitions of all of the important terms used in introducing genetics. More importantly it provides a thematic and integrated approach to the subject. The theme of the book is that genetics is concerned with the **transmission**, the **structure** and the **action** of the material in the nucleus (or prokaryote and virus) that is responsible for heredity. Wherever possible we have tried to break down the traditional barriers and to integrate the classical parts of the subject with the more recent developments at the molecular level. In particular we have drawn together heredity at the level of the chromosome and the DNA molecule. We have related the molecular basis of gene action and mutation to character differences and have discussed the division of the nucleus at mitosis and meiosis in terms of reproduction and life cycles. In Chapter 20 we have explained how genes are transmitted at the level of the population and how barriers to the free exchange of genes between different breeding groups can lead to the origin of new species. We have not gone into the realms of evolution above the species level.

We have also tried to confine the amount of material to a minimum and to avoid overloading the student with an excess of information which is either too detailed or simply non-essential. At the same time we have had to include some sections which are not generally found in 'A' level syllabuses—continuous variation, bacteria and viruses, cytoplasmic inheritance and genetic engineering. This has been done for the sake of completeness and to anticipate future developments in the curriculum. We also cherish the hope that some pupils will find the subject interesting for its own sake, and not simply as a means to an end.

To ease the burden on the reader we have organised the book in such a way that certain more difficult, or esoteric, parts have been separated off into boxed sections. The text is also subdivided into chapters which form topics or units that are as natural and self-contained as possible. Other features contributing to readability are the use of cross-referencing and the dilution of the text with numerous illustrations—many of which are originals and appear for the first time in this book.

Acknowledgements

Our task has been more difficult than we envisaged at the outset. We have been particularly concerned at the volume of material that is now required at this level, and the searching and demanding nature of many of the problems and questions that are set in 'A' level examinations. This is a problem which will grow as the explosion of knowledge goes on at an ever-increasing pace. It is one which the examiners will eventually have to deal with. For our part, we have been fortunate to have the helpful criticism and advice of several trusted friends and colleagues. In particular we would like to thank Dr John Parker, of Queen Mary College, for reading all of the chapters and for skilfully guiding us through some passages with which we were unfamiliar. Dr Mike Young, of the Department of Botany and Microbiology at Aberystwyth, gave us the benefit of his wisdom on the sections concerned with biochemical and microbial genetics. Dr Ben Miflin, of Rothamsted's Biochemistry Department, read Chapters 17 and 22 and his counsel was much appreciated.

We also record our thanks to Don Mackean who also read the entire manuscript in its final stages and made many constructive suggestions for simplifying parts of the text and for improving its accuracy and readability. Harvey Johnson, Science Editor at John Murray, gave valuable advice and encouragement during the 'embryogenesis' of the book, and we are grateful for the way in which he sustained us in the beginning.

We would also like to acknowledge the willingness of various examining boards in allowing us to reproduce questions from past 'A' and 'S' level papers. The boards concerned, together with abbreviations used in the text, are as follows:

Cambridge Local Examinations Syndicate (*C.*); Joint Matriculation Board (*J.M.B.*); London School Examinations Department (*L.*); Oxford Delegacy of Local Examinations (*O.L.E.*); Southern Universities Joint Board (*S.U.J.B.*); The Associated Examining Board (*A.E.B.*); Welsh Joint Education Committee (*W.J.E.C.*).

RNJ   AK

Contents

1
What is genetics?

The word 'genetics' was coined by William Bateson in 1907. He used it to describe a new branch of biology which began in 1900, after the rediscovery of Mendel's work on hybridisation in garden peas. **Genetics** *is the science of heredity.*

Heredity *is the process that brings about the similarity between parents and their offspring*; and Mendel had discovered a fundamental law, or rule, about how this process worked (Chapter 4). By similarity we mean that when plants and animals reproduce they have progeny of their own species, and not of some other kind. When human beings have offspring they are humans and not chimpanzees, or rabbits or any other organism. 'Like begets like'.

The members of a family are all *similar* to one another, and to their parents, in their characters (specific characteristics), but they also *vary* in many minor ways in the details of their individual development and appearance. In the human population of more than 4 billion people, each one of us can be uniquely recognised and distinguished from all of the others. These differences between the individuals of a family, or of a species, we refer to as **variation**. When we study genetics therefore we want to know how heredity can account for the *differences* between individuals as well as for their *similarities.*

Heredity and environment

It is important at the outset to realise that variation has two causes. The differences that we observe between individuals are only due in part to the internal factors of the cells that cause heredity. They are also partly accounted for by the external influences of the environment. In our own species (*Homo sapiens*) it is quite easy to find examples of inherited variation if we look at groups of people from widely separated parts of the world. The individuals shown in Fig. 1.1 all have distinctive physical features which enable us to recognise the groups to which they belong. The variations include differences in the colour of the skin and hair, texture of the hair, height, shape of the head and facial features. We know that these differences are largely due to heredity because they are known to have been passed on in the same form for several centuries. Moreover, when people migrate and settle in different parts of the world their descendants retain their racial characteristics regardless of the environments in which they live. On a lesser scale we can also see the way in which particular characters (colour of hair, skin and eyes) run in families within our own ethnic groups.

It is also obvious to us that not all differences between people are inherited. There are variations that arise due to the level of nutrition, others that are the result of exercise (i.e. large muscles) and some that may simply be due to accident or to a whole variety of other environmental causes. There is no doubt

(c) (d)

(e) (f)

Fig. 1.1 Heritable variation in man due to racial differences. (a) Chinese, from Hong Kong; (b) Masai, from Kenya; (c) West Indian; (d) European, from Switzerland; (e) Indian; (f) Thailander. (Photographs courtesy of Helene Rogers.)

in our minds that a person who lost a leg by accident would be able to have children who are perfectly normal with two legs! Likewise an olympic weight-lifting champion, who has built up his muscular physique by exercise and a high protein diet, is no more likely to have muscular offspring than a person who is deprived of good food and who is employed in making feather dusters. The development and appearance of a human being, or a buttercup, or a blue tit or any other species, is the outcome of influences due to both heredity and environment.

In some cases it is impossible to say what contribution is made to a particular character by heredity and what part is due to environment. This is particularly true where the differences between individuals are very small and where the character shows a continuous range of variation within a population (e.g. intelligence in man). In other cases where the differences between individuals are very clear-cut (e.g. dwarfism, or the presence of an extra finger in man), we can show that heredity is the cause of the variation by the way in which the character is passed on within a family over successive generations.

Reproduction

Sex cells

The key to understanding genetics lies in knowing what is inherited. What exactly is it that is passed on from the parents to their offspring during reproduction? The answer lies in the **sex cells**. Each parent contributes a single gamete to each of its progeny. The male donates the sperm and the female the egg. At fertilisation the gametes fuse to produce the fertilised egg, or zygote, from which the new individual develops.

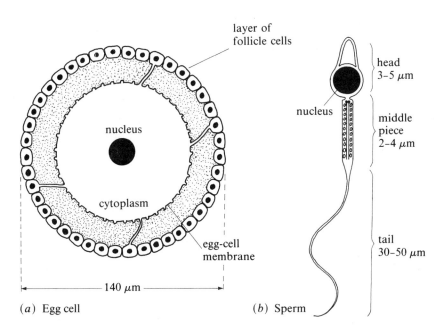

Fig. 1.2 Simplified diagram of the gametes of man.

Whatever it is that is 'passed on' must be contained within these two tiny cells. They provide the bridge between one generation and the next. A **gamete** is defined as *a mature reproductive cell which is capable of fusing with a similar cell of opposite sex to give a zygote.* A **zygote** is *a cell formed by the fusion of two gametes.* The sex cells of man are shown in Fig. 1.2. We will look upon them as representative of the gametes produced by all sexually reproducing organisms. Their fusion into a zygote constitutes the most significant single event in all our lives.

In studying genetics it is evident that we will have to look closely at the sperms and eggs. They are the vehicles of heredity. In sexual reproduction it is they alone that carry the genetic programs that determine the similarity of the offspring to their parents, and the heritable differences between one offspring and another. It is during their formation, in the male and female reproductive organs, that the *processes* which give rise to heredity and variation must take place. In the forthcoming chapters we will see what these processes are, and how we find out about them.

Somatic cells

In a multicellular organism the individual cells which comprise the **soma** (or body), i.e. the **somatic cells**, all arise by division from the single-celled zygote. These **cells** are *the units of structure and function* which comprise all living organisms. The general features of the cell are described in Box 1.1. During the multiplication of the body cells, to give a new organism, the processes of heredity must also take place. This has to happen because it is from the somatic cells that the sex organs will be formed, and eventually the next generation of sperms and eggs. In sexual reproduction the continuity of life proceeds through the gametes → zygote → somatic cell divisions → adult organism → new generation of gametes. All organisms, except for some unicellular ones, have to go through this phase of somatic cell division during their development. When the somatic cells divide, their genetic programs are passed on as an exact copy from one cell to another, so that all the cells of the body contain the same complete set of genetic instructions. What makes one body cell different from another one, in its form and function, is the way in which different parts of the genetic program are used in different cells; not that some cells have different programs to others. We know that all cells of the body carry a complete set of genetic instructions, because of what happens when some organisms *reproduce without sex*, i.e. **asexually**. This is well known in plants which can propagate vegetatively (Chapter 18). A small part of a plant can be detached, as in a geranium cutting, and this part can then develop into a new individual. The piece that is detached clearly contains a complete 'geranium genetic program' in each of its cells. If numerous cuttings are taken from the same parent it will also be observed that they are all genetically identical. The only differences that can be seen between them are those due to the environment. Some plants for example may receive less light than others and grow more slowly. They will appear different because of the environment, but there are no heritable variations between them.

We shall therefore have to look closely at somatic cell division. We need to know how it differs from the divisions that give rise to the sex cells, and how it provides for heredity without giving rise to any heritable variation.

Box 1.1 The plant cell

The cell is the unit of structure and reproduction in all living organisms. There is a wide diversity of types but in all species it conforms to the same basic plan. The substance which makes up the cell as a whole is the **protoplasm**, enclosed in plants within the cell wall. The protoplasm is further differentiated into the **nucleus** and the **cytoplasm**. In the cytoplasm are several kinds of organelles and other structures. These can be seen in Fig. 1.3, which shows the general features of a plant cell viewed under the electron microscope. Plant cells have two main features which are not present in animals, namely plastids and a rigid cellulose cell wall (bacterial cells are dealt with separately in Chapter 12).

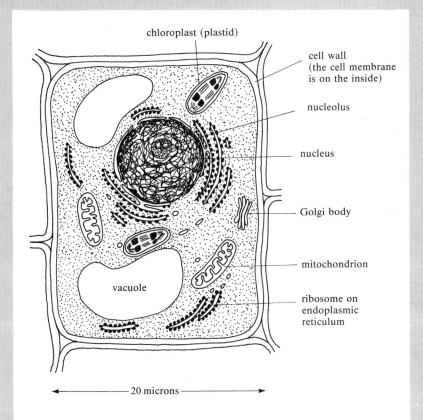

The part of the cell in which we are most interested is the nucleus. This is the organelle containing the genetic material which is responsible for heredity and variation. We will also be concerned, to a much lesser extent, with chloroplasts and mitochondria. These are organelles that have their own genetic information and are able to make more copies of themselves without instructions from the nucleus (Chapter 9). Another feature to note is the ribosomes. These numerous small bodies play a key part in the action of the genetic information. They are often associated with the endoplasmic reticulum.

The nucleus

The fertilised egg from which a new individual grows and develops must contain an enormous amount of information in its genetic program. In the case of man the inherited program of instructions in the cells must be capable of specifying all the details of the structure, the organisation and the functioning of a fully developed person of some 10^{13} cells. Instructions are required to determine the sex of an individual, the details of all the anatomy and physiology of the body, the complexities of the central nervous system and all the characteristics of our form and pigmentation (skin, hair, eye colour). Each species of living organism has its own unique set of instructions, or **genetic information** within its cells, to determine its own particular pattern of development.

In the study of genetics one of the first questions we want to answer is—where is this information located within the cell? Is it in the nucleus or in the cytoplasm, or in both? We suspect that it may be in the nucleus. The reason for thinking this is that the sperm and egg make an equal contribution to heredity (as will be seen later), yet the sperm contributes far less cytoplasm to the zygote than does the egg (Fig. 1.2). This simple question about the location of genetic information within the cell proved a difficult one to answer. Hämmerling made the first decisive experiment, using nuclear transplantation, in a single-celled alga called *Acetabularia*. His experiment is described in Fig. 1.4.

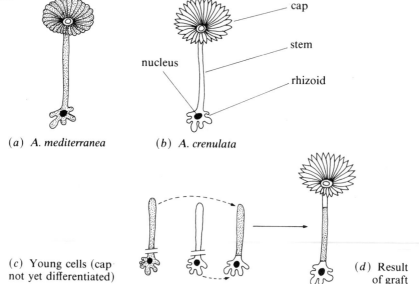

(a) A. mediterranea   (b) A. crenulata

(c) Young cells (cap not yet differentiated)   (d) Result of graft

Fig. 1.4 *Acetabularia* is a single-celled alga of tropical seas. It grows to about 6 cm in height and is differentiated into a rhizoid, a stem and a cap. The nucleus is located in the rhizoid. There are two forms, A. *mediterranea* (a) and A. *crenulata* (b), which differ in cap morphology. Hämmerling performed a grafting experiment which showed that the genetic information which determined the difference between the two species resided in the nucleus, rather than in the cytoplasm. Young cells were cut in two, and the rhizoid (with nucleus) of the *crenulata* type was grafted onto the stem portion of *mediterranea* (c) and vice versa. The cap which developed was characteristic of the species contributing the nucleus (d).

In later chapters we will give much more evidence to confirm that the genetic information is located mainly in the nucleus. We will also point out that some of it is contained in other cell organelles—namely the chloroplasts and mitochondria. Organisms which do not have their genetic material enclosed within a nucleus are dealt with separately in Chapter 12.

Genetics

In studying the science of heredity geneticists are principally concerned with three questions about the information contained within the nucleus: its **transmission**, its **structure** and its **action**.

Transmission

How is the genetic information handed on in an exact copy when somatic cells divide? How is it passed on during sexual reproduction so that all the offspring resemble their parents yet differ uniquely from one another? How is it distributed in natural populations so that it may change in response to the forces of selection and evolution?

Structure and organisation

What is the chemistry of genetic information? What kind of material carries the information and what is its molecular structure and organisation within the nucleus?

Action

How does the genetic material act within the cell to determine the characters of living organisms?

In this book we will answer these questions—as far as the answers are known. We will begin by dealing with transmission at the level of the cell and the individual organism (Chapters 2-12). Then we will consider the structure, the organisation and the action of the genetic material (Chapters 13-17). Chapters 18-20 will deal with transmission at the level of the population and between species during evolution. In the last two chapters we will consider aspects of the experimental manipulation and engineering of genetic material. Genetic engineering offers prospects of improving food production and more effective ways of overcoming disease.

Summary

Genetics is the science of heredity. Geneticists study the transmission, the structure and the action of the material in the cell which is responsible for heredity. These studies are made at all levels of organisation, from molecules and single cells through to individuals and groups of individuals comprising populations.

2
Mitosis and meiosis

The genetic material responsible for heredity and variation is located mainly in the cell nucleus. One way to study genetics therefore is to observe what happens to the nucleus when a cell divides. There are two kinds of nuclear division that we have to study—mitosis and meiosis.

Mitosis

Mitosis is *division of the nucleus into two daughter nuclei that are genetically identical to one another and to their parent nucleus.* It takes place in somatic cells, during development, and is part of the process of cell division. The way that it happens is essentially the same in all **eukaryotes**, i.e. in all organisms which have a true nucleus.

To study mitosis we observe nuclei in cells which are actively dividing. In plants this means looking at the meristematic regions (growing points) of the roots and shoots. In animals there are several sources of material. Insects can be studied by squashing out young embryos; while in man, and other mammals, it is now standard practice to use cell cultures, e.g. white blood cells grown in a culture medium.

When we look at a mass of cells, in a piece of squashed-out tissue, we see that they are usually not synchronised in their division, and only a small proportion of them are in mitosis at any one time. The bulk of the nuclei are in **interphase,** i.e. in *the stage in which they are not visibly engaged in division.*

The interphase nucleus

The form and appearance of the interphase nucleus can be seen in Figs 2.1 and 2.2. At this stage it is a spherical structure bounded by a nuclear membrane and containing a mass of chromatin, nuclear sap and one or more spherical nucleoli. The membrane is a double structure and is perforated by a number of pores which allow for the movement of certain materials between the nucleus and the cytoplasm. Interphase nuclei have a range of sizes. One which is ready to divide contains twice the amount of chromatin, and is about double the size, of one which has just completed division. The **chromatin** component is a complex substance made up of approximately equal proportions of deoxyribonucleic acid (**DNA**, which is the genetic material) and protein. Under the light microscope chromatin has a granular appearance and we are unable to see the details of its structure. The electron microscope shows that it consists of a mass of very thin bumpy fibres with a diameter of about 250 Å ($=25$ nanometres $= 2.5 \times 10^{-8}$ m). The **nuclear sap** is a fluid which is rich in enzymes and various other kinds of molecules concerned with the *activity* of the genetic material (Chapters 13 and 14). The **nucleoli** are spherical bodies which are

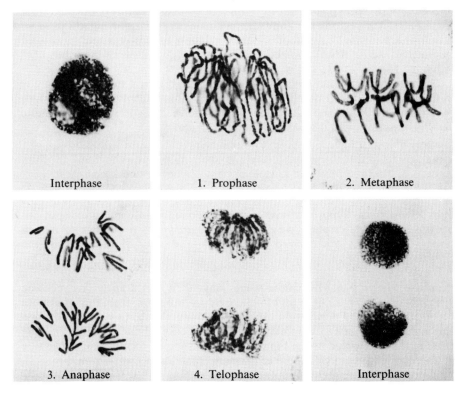

Fig. 2.1 Photographs of mitosis in root meristem cells of the onion (*Allium cepa*). *A. cepa* is a diploid species with 8 pairs of chromosomes ($2n = 2x = 16$) in its somatic cell nuclei. Interphase nuclei, as they appear just before and just after mitosis, are also shown.

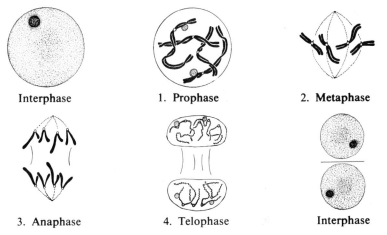

Fig. 2.2 Diagram showing mitosis in an organism with two pairs of chromosomes. The interphase nuclei are also shown as they appear just before and just after mitosis. Cell structures other than the nucleus are not shown in the diagrams. The small spherical bodies in the interphase nuclei, and attached to one pair of chromosomes in prophase and telophase, are nucleoli.

formed around specialised regions of one (or sometimes more) of the chromosome pairs. The regions are known as **nucleolus organisers**. Nucleoli have no membrane and they consist of an accumulation of proteins and ribosomal ribonucleic acid (**RNA**) molecules which will give rise to the ribosomes (Chapter 14).

The interphase nucleus is often misleadingly referred to as being in the 'resting stage' of the division cycle. This is because it shows no visible signs of division at this time, but it does in fact have a high level of metabolic activity which we will discuss in Chapters 13 and 14.

Mitosis

The nucleus undergoes a striking sequence of changes in its form and organisation when it divides during mitosis (Figs 2.1 and 2.2). We know from cine films made from living material that the sequence is continuous, but for convenience of description we divide the process into a number of named stages. The names are derived from Greek and refer to the form of the chromosomes.

Prophase   In prophase the **chromosomes** make their appearance. They become visible because they shorten and thicken, by spiralisation, from the interphase chromatin fibres. Each chromosome is thread-like and consists of two **chromatids** (the **sister chromatids**) running throughout its length. The chromatids are twisted around one another and joined together on either side of an uncoiled region called the **centromere** (or **primary constriction**, Fig. 2.3).

centromere

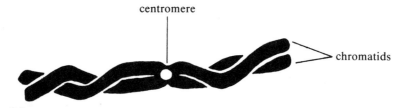

chromatids

Fig. 2.3

Another uncoiled region can be seen at the site of attachment of the nucleolus. This is the **nucleolus organiser** or **secondary constriction** (it is best seen in the metaphase of c-mitosis, Fig. 2.5). As prophase proceeds the chromosomes become progressively shorter and thicker, and appear as more distinct and separate structures. When they reach their maximum degree of contraction the nuclear membrane breaks down and this marks the end of prophase.

Pro-metaphase   The main event in pro-metaphase is the formation of the **spindle**. As its name implies this is an ellipsoidal, or barrel-shaped, structure which is organised in the cytoplasm of the cell. It is broadest at the **equator** and tapers off towards the two extremities, or poles. The spindle is composed of **spindle fibres,** or **microtubules**. These are long thin cylindrical filaments assembled from a protein called **tubulin**. Some of the fibres stretch from pole to pole and others run from the poles to the centromeres of the chromosomes. When the spindle is organised the chromosomes attach to its fibres at random places by their centromeres. The centromeres then move to the equator of the spindle.

Metaphase   At metaphase the centromeres are held at the equator of the spindle, under the pull of their spindle fibres, in a position which is equidistant from the two poles. At this stage, as in prophase, the sister chromatids lie closely parallel to one another throughout their length.

Anaphase   The two parts of the double centromere, i.e. the half-centromeres, begin to separate from one another. The attachment of the sister chromatids, in the regions on either side of the centromeres, then lapses (i.e. they cease to be held together) and the chromatids are peeled apart as their centromeres are drawn to opposite poles. Once separated from one another the sister chromatids are referred to as **daughter chromosomes**.

Telophase   In the final stage of mitosis the centromeres are drawn together at the poles of the spindle and the nuclear membrane is reformed around the two groups of daughter chromosomes. The prophase coiling sequence is reversed and the daughter chromosomes go back into a diffuse interphase form. Nucleoli also reappear at the sites of their organiser regions. Mitosis is thus complete and we have two daughter nuclei each containing an identical set of single-stranded chromosomes in place of the double-stranded ones which were present in the parent nucleus.

Cytokinesis

After mitosis has ended **cytokinesis**, or division of the cytoplasm, usually takes place. It is the second part of the cell division process and it separates the cytoplasm, and the newly formed daughter nuclei, into two daughter cells. The mechanism of cytokinesis differs in plant and animal cells. Animal cells undergo **cleavage** by constriction of the cytoplasm and furrowing of the cell membrane. In plants a cell plate forms across the equator.

Mitotic cycle

The term **mitotic cycle** (or **cell cycle**) refers to *the life cycle of an individual cell.* It includes all of the events which take place between a given stage on one division cycle around to the corresponding stage in the next cycle. We are using the term 'mitotic cycle' here with reference to events going on within the nucleus. The way in which the cycle is divided up, in relation to the division and reproduction of the chromosomes, is shown as a diagram in Fig. 2.4.

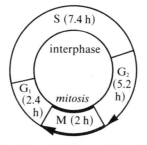

Fig. 2.4   Diagram of the mitotic cycle. Times given for the various stages are for meristematic root tip cells of the onion (*A. cepa*), at 20 °C. M = mitosis (prophase → telophase); G_1 = gap 1; S = DNA synthesis and chromosome duplication; G_2 = gap 2.

We know, from various kinds of timing experiments, that mitosis itself occupies only a small part of the total mitotic cycle. The bulk of the period is spent in interphase. In the meristematic root tip cells of the onion for example, mitosis (M) takes an average of two hours. Interphase occupies 15 h, and this phase is further subdivided into G1 (Gap 1), S (period of DNA synthesis and chromosome duplication) and G2 (Gap 2). The chromosomes duplicate themselves, and become double-stranded, during the mid part of interphase.

Details of the molecular events involved in DNA synthesis, and in chromosome structure and duplication, are dealt with in Chapter 13. The duration of the mitotic cycle varies from one species to another. In general its length is related to the amount of genetic material contained within the nucleus. Larger nuclei have longer cycles. What happens to the nucleus once a cell stops dividing is explained in Box 2.1: most of them end up in G1.

Box 2.1   Endomitosis

At some stage during development most cells stop dividing and become differentiated into the permanent tissues of the organism. When this happens they are usually arrested in the G1 stage of the mitotic cycle with their interphase chromosomes in the single-stranded condition. **Karyogamy** (division of the nucleus) and cytokinesis are thus coupled, and both cease their activities in the same division cycle. In some cells, however, this is not the case. There is a modified form of mitosis—called **endomitosis**—in which the chromosomes continue to duplicate for several more cycles after cell division has stopped. There is no spindle formation and the duplication takes place at interphase within the nuclear membrane. It has two main forms: **endopolyploidy** and **polyteny**. In endopolyploidy the chromosomes duplicate and separate repeatedly so that their number increases through a doubling series. Endopolyploid cells can be recognised by the large size of their interphase nuclei and they are commonly found in many of the differentiated tissues of both plants and animals. They are well known for instance, in the liver cells of man and in the tapetal cell layer of the anthers of some flowering plants. Polyteny is less common. In this process the interphase chromosomes replicate themselves without separating at the centromeres, and they become progressively more and more multistranded. Eventually they may come to consist of several hundred, or even several thousand, chromatids—depending upon the number of cycles of **endoreduplication**. Polytene nuclei are best known in the salivary gland cells of larvae of dipteran flies, such as *Drosophila* and *Chironomus*. They are also found in some plant tissues, such as the suspensor cells of the *Phaseolus* embryo. Their importance in relation to development is discussed in Chapter 17.

C-Mitosis

The chromosomes are the carriers of the genetic information. It is therefore of some interest to study their form and to see how they vary in number and size among a wide diversity of species. As it happens, the normal form of the mitotic division is not ideally suited to this purpose. This is because in their most contracted state the chromosomes are tightly grouped together around the equator of the spindle and it is difficult to see, and to analyse, them individually. To overcome this inconvenience we pretreat the living cells, before fixing and staining them, with a spindle-inhibiting drug called **colchicine**.

This substance interferes with the arrangement of the protein microtubules (spindle fibres) and breaks down the structure of the spindle. It also prevents the formation of any new spindles as long as it is being applied to the cells. As a consequence of this treatment, the mitotic cycle is blocked at the stage where the chromosomes are in their metaphase form, and because there is no spindle they lie freely dispersed throughout the cytoplasm. This condition is known as **c-mitosis** (colchicine mitosis) and it is an ideal experimental method for the study of metaphase chromosomes. An added advantage is that the chromosomes become more contracted than they are in a natural metaphase. The same effect can be produced by a number of other spindle poisons (e.g. alpha-bromonaphthalene). A c-mitosis from the onion is shown as one of the examples in Fig. 2.5. Compare the form of the chromosomes in this photograph with those in the normal metaphase in Fig. 2.1. In the study of metaphase chromosomes we refer to a set which has been classified, according to their number, their form and their size, as the **karyotype**. The karyotype of man is shown as an example, in Fig. 2.6. In constructing a karyotype the position of the centromere (and that of the secondary constriction) can provide a useful marker for chromosome classification. The location of the centromere is fixed for any one chromosome, but it varies for different members of the set. We recognise three main categories of chromosomes with respect to their centromere location: (1) **metacentric** = centromere approximately in the middle; (2) **acrocentric** = centromere near to one end so that one arm is small and the other one much longer; (3) **telocentric** = terminal centromere (Fig. 2.7).

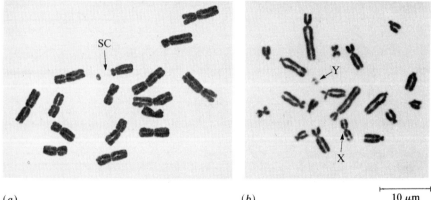

(a)   (b)   10 μm

Fig. 2.5   Metaphase of c-mitosis in (a) the onion (*Allium cepa* var. 'viviparum') and (b) the red kangaroo (*Macropus rufus*). The onion has 8 pairs of chromosomes which are mostly metacentric (centromeres near the middle). Note the large secondary constriction (SC) in one of the chromosomes. In the red kangaroo the 10 pairs of chromosomes are mainly acrocentric (centromeres near the end). The cell shown here is from a male and has an XY sex chromosome pair (Chapter 7). (Photograph b kindly supplied by Dr David Hayman, Genetics Department, University of Adelaide.)

Chromosome pairs

In the study of mitosis, and of c-mitosis, we see that the chromosomes are usually present in the cell in identical pairs—which we call **homologues**. The identity of the homologues is most obvious in those species in which the pairs

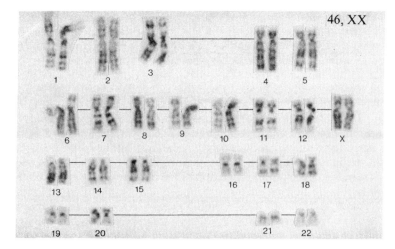

(a) Karyotype of a normal female

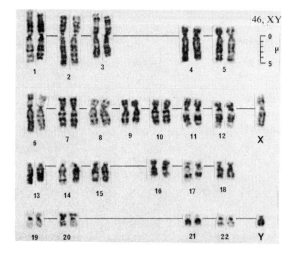

(b) Karyotype of a normal male

Fig. 2.6 Karyotypes of man showing the chromosome sets of (a) the female (A.K.) and (b) the male. The difference between the sexes is due to the sex chromosomes (XX♀, XY♂), which are dealt with in Chapter 7. To construct the karyotypes, photographs are first taken of c-mitosis in white blood cells growing in culture. The individual chromosomes are then cut out and classified into pairs on the basis of their size and form.

The classification is greatly aided by the use of a special staining procedure which gives the chromosomes a banded appearance. This banding is related to the underlying molecular organisation of the chromosomes and it allows us to identify all the individual chromosome pairs. (Photographs by courtesy of Dr David Hughes, Institute of Child Health, University of London.)

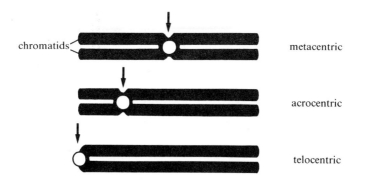

Fig. 2.7 The three main forms of metaphase chromosomes classified according to position of the centromere (arrowed).

are of differing sizes and differing centromere positions (Fig 2.5*b*, 2.6). *Cells, phases of the life cycle, and organisms which are characterised by having two sets of chromosomes in their cells* are said to be **diploid**. The organisms with which we are mainly concerned in the first part of this book are all diploid for the major part of their life-cycle, as explained in Chapter 3.

Chromosome numbers

As we have already seen, chromosome numbers vary from species to species. To write down chromosome numbers we use a convention whereby:

x = the **basic number** of different chromosomes in a single (**haploid**) set;

n = the **gametic number**, or number in the gametes;

$2n$ = the **zygotic number** found in the zygote and in the somatic cells derived from it.

In diploids the gametic number (n) and the basic number (x) correspond, so that $n = x$. In the onion (*Allium cepa*) we write the chromosome number as $2n = 2x = 16$, indicating that the species is diploid with two sets of chromosomes in its somatic cells. The chromosome number of the broad bean (*Vicia faba*) we write as $2n = 2x = 12$ and that of the crocus (*Crocus balansae*) as $2n = 2x = 6$. It is necessary to use this system because some species, especially among flowering plants, have more than two sets of chromosomes, i.e. they are **polyploid**—$3x$, $4x$ etc. as discussed in Chapter 16. In animals where diploidy is the general rule it is usual to give the chromosome number in an abbreviated form; e.g. in the red kangaroo (*Macropus rufus*) $2n = 20$, in man $2n = 46$ and in the fruit fly (*Drosophila melanogaster*) $2n = 8$.

Mitosis and heredity

What does the study of mitosis tell us about the process of heredity?

1. The first thing we see is that the genetic material in the nucleus is divided up into a number of thread-like bodies which we call the **chromosomes**.
2. The chromosomes of a set are not all alike. They vary in their size and in the positions of their centromeres and secondary constrictions.

3. Chromosomes reproduce themselves in interphase into two identical sister chromatids.
4. During mitosis, chromatids separate so that one sister chromatid from each chromosome goes into each of the daughter nuclei. Each daughter nucleus therefore contains an identical set of chromosomes.

Heredity in somatic cells is determined by the accuracy of chromosome reproduction and by the continuity of the structure of each chromosome from one cell division to the next. Mitosis is a mechanism for distributing identical copies of the genetic information to daughter cells during the process of growth and development.

In terms of a single chromosome in mitosis the process of heredity can be simply summarised as shown in Fig.2.8.

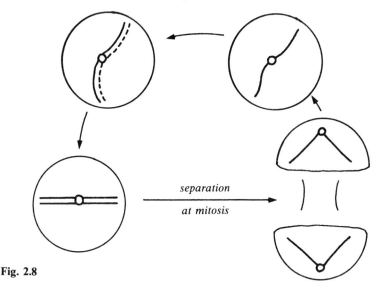

separation

at mitosis

Fig. 2.8

Meiosis

Meiosis is *the reduction division of the nucleus in which the zygotic (diploid) number of chromosomes is reduced to the gametic (haploid) number.* This means that in a diploid organism each cell that results from meiosis ends up with only one of each pair of chromosomes, instead of two as in mitosis. In animals, meiosis gives rise directly to the gametes, but in plants it produces haploid **spores** from which the gametes later arise by mitosis (Chapter 3). When two haploid gametes come together at fertilisation, the diploid number of chromosomes is restored again in the zygote. The regular alternation of meiosis and fertilisation in the life cycle therefore maintains the constancy of the chromosome number from one generation to the next. The way in which meiosis fits into the life cycles of different groups of organisms, relative to fertilisation, is discussed in the next chapter.

To find meiosis we look inside the anthers and ovules of flowering plants, and in the testes and ovaries of animals. We usually work with the male reproductive organs (anthers or testes) because they have a much greater abundance of cells in meiosis than those of the female (Chapter 3).

Visible events of meiosis

The events which take place during meiosis are shown in the sequence of photographs in Fig. 2.10, which are from anther squashes in rye (*Secale cereale*), and in the diagrams in Fig. 2.9. As we see, meiosis is more complex than mitosis. It involves two divisions of the nucleus, but the chromosomes only reproduce once. The stages of each of the two divisions are named in the same way as in mitosis, except that the first prophase is very much longer and is divided into five sub-stages:

Division I ⟶	Interphase	⟶ Division II
Prophase I (*a*) Leptotene (*b*) Zygotene (*c*) Pachytene (*d*) Diplotene (*e*) Diakinesis Metaphase I (MI) Anaphase I (AI) Telophase I		Prophase II Metaphase II (MII) Anaphase II (AII) Telophase II

Although meiosis proceeds in virtually the same way in all species which have a nucleus, the individual stages can be seen much better in some of them than in others. Details of the early prophase stages (*a, b, c*) are shown separately in the photographs from the lily (*Lilium* 'Enchantment') in Fig. 2.11; and those of diplotene in the photograph from the grasshopper (*Chorthippus parallelus*) in Fig. 2.12.

Meiosis is always preceded by an interphase stage during which the genetic material of the nucleus is duplicated in readiness for division. When division begins the chromosomes make their appearance as they shorten and thicken by coiling of their long interphase fibres.

Prophase I

(1) Leptotene   The chromosomes become visible as **single threads**. Although the genetic material has already been duplicated (as we know by measuring its amount) this is not evident by light microscopy of the chromosomes. In this respect the early prophase of meiosis differs from that of mitosis.

(2) Zygotene   Pairing (or **synapsis**) takes place between homologous partners of each pair of chromosomes. It may begin at one or more sites along the chromosomes, and as it proceeds it brings them together in close parallel alignment throughout their length. The pairing is seen to be precise and to bring corresponding parts of the two homologues into intimate contact.

(3) Pachytene   The homologues are now fully paired and contracted into a much shorter and thicker form than in zygotene. **Chromomeres**, bead-like structures due to regions of localised coiling along the chromosomes, are clearly visible at this stage. During pachytene the chromosomes become double-

Leptotene

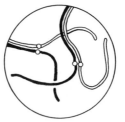
Zygotene

Pachytene

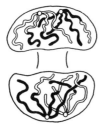
Diplotene/Diakinesis

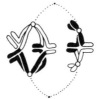
Metaphase I

Anaphase I

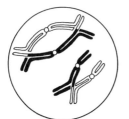
Telophase I

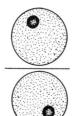
Interphase

Prophase II

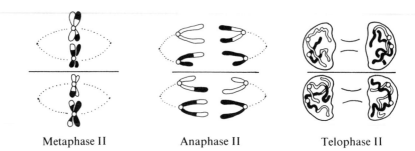

Metaphase II   Anaphase II   Telophase II

Fig. 2.9   Diagram of the stages of meiosis in an organism with two pairs of chromosomes. Cell walls are not shown.

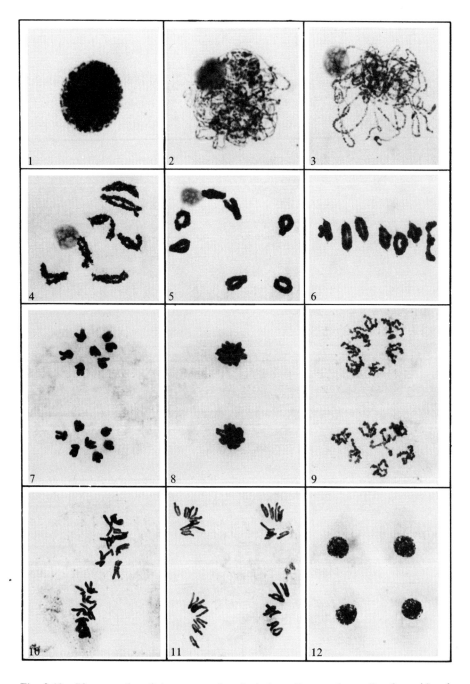

Fig. 2.10 Photographs of the stages of meiosis in pollen mother cells of rye (*Secale cereale*, $2n = 2x = 14$). (1) Interphase/early leptotene. (2) Zygotene. (3) Pachytene. (4) Diplotene. (5) Diakinesis. (6) Metaphase I. (7) Anaphase I. (8) Telophase I. (9) Prophase II. (10) Metaphase II. (11) Anaphase II. (12) Tetrad.

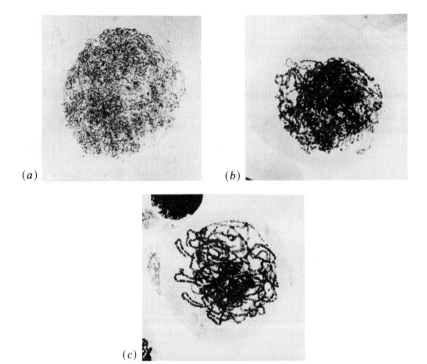

Fig. 2.11   Early prophase I stages of meiosis from anther squashes in the flowering plant *Lilium* 'Enchantment': (*a*) leptotene, (*b*) zygotene, (*c*) pachytene. (Photographs kindly supplied by Dr John Parker, Queen Mary College, University of London.)

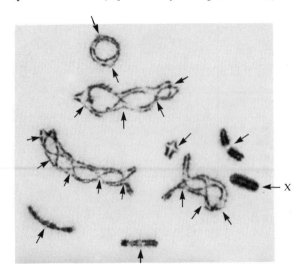

Fig. 2.12   Diplotene of meiosis in the grasshopper *Chorthippus parallelus* ($2n = 16 + X$). The photograph is from a testis squash of a male. The chromatids and the chiasmata (arrowed) can be clearly seen, as well as the single X chromosome. Three of the small bivalents each have a single terminal chiasma. (Photograph by courtesy of Professor Bernard John, Research School of Biological Sciences, Australian National University.)

stranded and form **chiasmata**. We cannot see these two events happening at this stage, because of the close and intimate association between the homologues, but the evidence for their occurrence is revealed in the next stage.

(4) Diplotene Homologues repel one another and begin to move apart at their centromeres. The force of attraction which held them together until the end of pachytene lapses, but the attraction between the sister chromatids remains. In each **bivalent** (pair of associated homologues) we now see the chromatids and the chiasmata. This is demonstrated most clearly in the diplotene cell of the grasshopper shown in Fig. 2.12—it is not always so obvious in other species. Each chiasma is the result of a single break in each of two *non-sister* chromatids, at exactly corresponding places, followed by crosswise rejoining of the broken ends in new combinations (Fig. 2.13). This *process of breakage and rejoining of chromatids, by which a visible chiasma is formed,* is known as **crossing over**. The term **chiasma** refers to *the cross-shaped arrangement of the chromatids which results from the crossing over.* As far as we can tell, the position of the chiasma corresponds with the point of exchange of the chromatids (see Box 2.2). When a second (or third) chiasma is formed in a bivalent it may involve the same pair, or another pair, of non-sister chromatids. It is because of the way in which the sister chromatids attract one another that the two homologues are held together to give the chiasma, following their breakage and rejoining. The chiasma thus serves a mechanical function which is important in holding the homologues together and in providing for their orderly separation in the subsequent stages. It also has a genetical consequence which is explained in the more detailed account of the chiasma given in Box. 2.2.

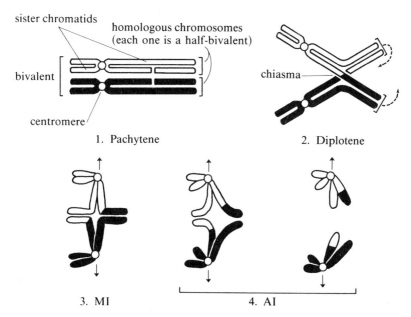

Fig. 2.13 Diagram showing how a single chiasma is formed, as a result of crossing over, and the way in which it gives rise to the MI bivalent and is then resolved when the attraction between the sister chromatids lapses at AI.

Box 2.2   Chiasmata

Chiasmata are a source of perpetual error and confusion. The first thing to understand, in trying to get the better of them, is the way in which they arise by crossing over in prophase I. Breaks occur at precisely corresponding places in the two non-sister chromatids, and these are then rejoined crosswise in new combinations to give a cross-shaped, or X-shaped, configuration at the point of exchange, as seen in diplotene. In drawing diagrams it is important to remember that the *sister chromatids* always remain in close association, from the time they are formed until their mutual attraction lapses in AI. The diagrams in Figs 2.13 and 2.14 illustrate the formation of chiasmata and the way in which their number and position determine the form of the MI bivalents.

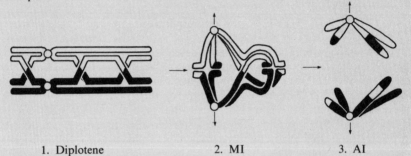

| 1. Diplotene | 2. MI | 3. AI |

Fig. 2.14   A complex bivalent showing three chiasmata at diplotene and their consequences at MI and AI. (For convenience all three chiasmata are drawn in the same pair of non-sister chromatids.)

It is not normally possible, of course, to see the breakage and rejoining events, that are depicted in Figs 2.13 and 2.14, by looking at chromosomes down the microscope. But in organisms where it has been possible to investigate these affairs, it turns out that crossing over does indeed take place at the four-strand stage, after chromosome duplication, and that the points of exchange correspond with the positions of the chiasmata.

The other important point to note about chiasmata is the way in which their formation leads to an exchange of genetic material between the two homologues which pair together to form the bivalents. The consequences of this exchange can be illustrated more clearly in the AI half-bivalents when the chiasmata have been resolved (Figs 2.13 and 2.14). It is because the half-bivalents are now composed of two chromatids which are genetically non-identical that the second of the two meiotic divisions is required—in order to separate them into different nuclei. This aspect is explained more fully in the section on meiosis and heredity.

(5) Diakinesis   The chromosomes undergo further shortening and thickening and begin to approach their maximum degree of contraction. The nucleolus disappears at this stage and the nuclear membrane also begins to break down. Prophase I is now at an end.

Metaphase I

The spindle is organised at pro-metaphase I (the stage before metaphase I) and the bivalents attach to it by their centromeres and move towards the equator. At metaphase I (MI) the homologous centromeres of each bivalent take up positions in which they are equidistant on opposite sides of the equator, and they show signs of being pulled in the direction of the poles.

Anaphase I

At anaphase I (AI) the attraction between sister chromatids lapses, and the half-bivalents are drawn to opposite poles by their centromeres. It is at this stage that the number of centromeres is reduced from the diploid to the haploid value.

Telophase I

The chromosomes uncoil and become diffuse again in their appearance. In some species they go into an interphase resting stage, but in others they do not. Where there is an interphase, it is different from that in mitosis because there is no duplication of the chromosomes. At the end of telophase I the nucleus has completed its first division.

Prophase II

A second cycle of coiling and shortening begins, and the chromosomes are visible again. They differ from those seen in prophase of mitosis in that their chromatids are widely-splayed, and do not lie in close parallel alignment. By this stage, of course, the chromatids are not sisters any more—they are genetically non-identical as a result of the crossing over that took place in pachytene.

MII, AII and Telophase II

These stages are similar to those already described for mitosis.

At the end of meiosis four haploid nuclei are produced. Each one contains a basic set of chromosomes. What happens to them next depends upon the kind of organism concerned, and whether they have been formed in the male or female reproductive organs. These matters are considered further in Chapter 3, which deals with life cycles and gametogenesis.

Meiosis and heredity

Meiosis provides for heredity. It does so by distributing a complete haploid set of chromosomes, and therefore a complete set of genetic information, into each of the gametes. The gametes in turn serve as the vehicles of heredity and transmit their chromosome sets to the zygotes which are formed at fertilisation. In this way heredity, or the process by which character determinants are handed on from parents to their offspring, takes place.

Meiosis also provides for variation. It does this because the sets of chromosomes which are distributed into the gametes are not necessarily identical to one another in the way in which they are following mitosis. *The products of meiosis are non-identical.* Precisely what we mean by 'non-identical' will be made much clearer in the later chapters. For the moment we simply have to appreciate that meiosis takes place in diploid cells which contain *two* sets of chromosomes—one set which came from the 'mother' (egg cell) and one set which came from the 'father' (sperm) at fertilisation. We refer to these as the **maternal** and **paternal** sets. Each set carries the full complement of genetic information, but in each set the information which specifies individual character differences (e.g. eye colour in man) may be present in a slightly different form (e.g. brown or blue). During prophase I of meiosis, the paternal and maternal chromosomes of each pair come together and exchange parts of their chromatids with one another by crossing over. This occurs in such a precise way that the recombined chromatids neither lose nor gain information, but simply carry new combinations of the different forms of information that were present in the maternal and paternal partners. The non-identical nature of the gametes also depends upon how the maternal and paternal chromatids from *different pairs* of homologues are combined together following the reduction division—as explained in Chapter 6.

In dealing with the stages of meiosis, and with its provision for heredity and variation, it is useful to have a simplified scheme which summarises the main features of the process. This summary is given in the form of Fig. 2.15.

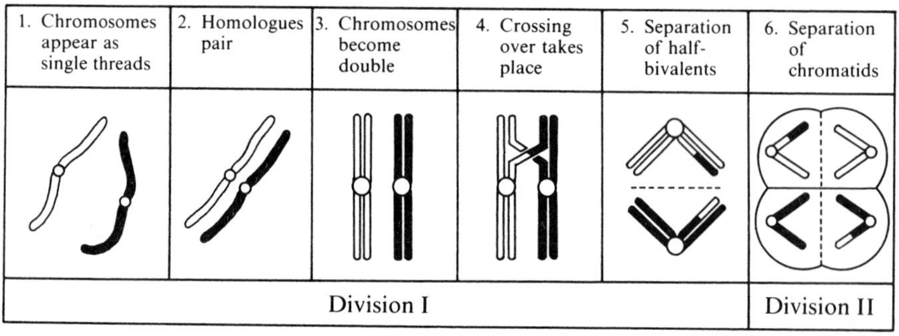

Fig. 2.15   Summary of the main events of meiosis (based on an idea by K. R. Lewis and B. John, in *The Matter of Mendelian Heredity*, Churchill 1964).

Differences between mitosis and meiosis

Mitosis	Meiosis
Nucleus divides once, chromosomes once	Nucleus divides twice, chromosomes once
Two daughter nuclei	Four daughter nuclei
Chromosome number constant	Chromosome number halved
Chromosomes are double when first seen	Chromosomes are single when first seen
Homologues independent	Homologues associate in pairs
Products genetically identical	Products genetically non-identical

Summary

There are two kinds of division of the nucleus: mitosis and meiosis. Observation of these divisions gives us information about the processes of heredity and variation. We reason that the genetic material in the nucleus is present as a number of thread-like bodies, called chromosomes, which maintain their organisation throughout the division cycles.

In the mitotic cycle the chromosomes are duplicated, during interphase, and then their chromatids are separated so that each daughter cell contains an identical set of daughter chromosomes. The cell products of mitosis are therefore genetically identical.

Meiosis is the reduction division which occurs in the sexual cycle of reproduction. It reduces the zygotic $(2n)$ number of chromosomes to the gametic (n) number. Meiosis provides for both heredity and variation. Heredity takes place because each of the cell products contains a complete set of the basic complement of chromosomes. Variation arises from the way in which the cells carry different mixtures of the maternal and paternal chromatids, as well as recombined chromatids which result from crossing over. The products of meiosis are therefore genetically non-identical.

Further reading

Dyer, A. F. (1979), *Investigating Chromosomes*, Edward Arnold.
John, B. and Lewis, K. R. (1972), *Somatic Cell Division*, Oxford Biology Reader No. 26, Oxford University Press.
John, B. and Lewis, K. R. (1984), *The Meiotic Mechanism*, Carolina Biology Reader No. 65, Carolina Biological Supply Company.
Lewis, K. R. and John, B. (1972), *The Matter of Mendelian Heredity*, 2nd edn, Longman.

Questions follow Chapter 3.

3
Life cycles

The nuclear divisions of mitosis and meiosis become more meaningful when considered in terms of the life cycles of growth and reproduction of different kinds of organisms. Mitosis is part of the cell division process which gives rise to growth and development, and as we will explain in this chapter it is involved in both the diploid and haploid phases of the life cycle. Meiosis is involved in sexual reproduction and in the reduction division which changes the diploid phase of the life cycle into the haploid one. Fertilisation is the complementary process which restores the diploid condition by combining together two haploid gametes into a zygote. **Sexual reproduction** is defined therefore as *the regular alternation in the life cycle of meiosis and fertilisation to provide for the production of offspring.* Since a large part of the study of genetics is concerned with parents and their offspring (Chapter 1) it is essential that we know something about how sexual reproduction takes place. In particular we need to know where meiosis and fertilisation fit into the life cycle, and how the gametes and zygotes are produced.

Life cycles

There are three main patterns of life cycle. They depend upon the duration of the diploid and haploid phases, and where meiosis occurs relative to fertilisation:

1. **haplontic**—haploid for most of the life cycle;
2. **haplo-diplontic**—alternating haploid and diploid phases;
3. **diplontic**—entirely diploid except for the gametes.

Haplontic cycles

In **haplontic cycles** *the organism is haploid for the major phase of the life cycle.* The adult organism is haploid and therefore produces gametes by mitosis rather than by meiosis. These unite during fertilisation to form a diploid zygotic cell which then undergoes meiosis resulting in the restoration of the haploid condition. The diploid phase is therefore restricted to a single cell, the zygote. Meiosis may occur immediately after fertilisation, or after a resting period. The cycle can be represented as in Fig. 3.1.

Examples are found in unicellular green algae, e.g. *Chlamydomonas*, the filamentous algae, e.g. *Spirogyra*, and many other algae. Haplontic cycles are also common in the fungi and in protozoa. The life cycle of the saprophytic fungus *Sordaria brevicollis* is given as a representative example (Fig. 3.3). *Sordaria* is a haploid organism ($n = 8$). The adult phase consists of a mycelium made up of a network of filamentous hyphae. Sexual reproduction only occurs

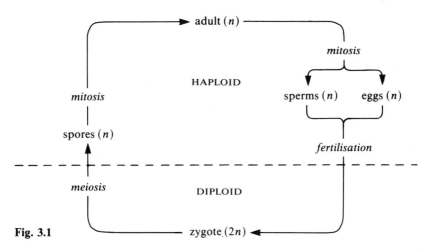

Fig. 3.1

when mycelium of opposite mating type strains is present, otherwise the fungus will keep growing asexually by making vegetative growth. The diploid phase is confined to a single cell called the ascus initial cell.

There is a wide diversity of reproductive forms within the fungi and not all of them have this kind of life cycle: some have no known sexual reproduction at all, i.e. the fungi imperfecti. *Sordaria* is a useful example because it is an organism which is widely used in genetic experimentation (Chapter 6).

Haplo-diplontic cycles (alternation of generations)

In haplo-diplontic cycles the organism exists in both haploid and diploid forms, both of which develop by mitosis. The haploid form, or **gametophyte**, produces gametes which unite to give the diploid zygote. The zygote divides mitotically to form the diploid spore-producing phase, or **sporophyte**, from which haploid spores are formed by meiosis. *The regular alternation of haploid and diploid phases in the life cycle of sexually reproducing plants* is known as the **alternation of generations** (Fig. 3.2).

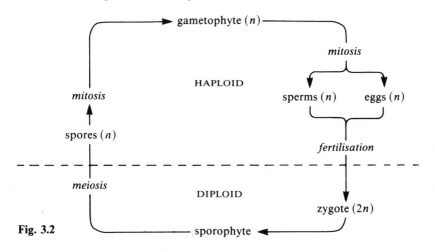

Fig. 3.2

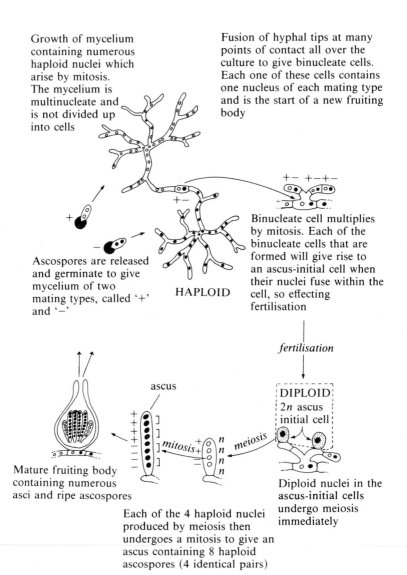

Growth of mycelium containing numerous haploid nuclei which arise by mitosis. The mycelium is multinucleate and is not divided up into cells

Fusion of hyphal tips at many points of contact all over the culture to give binucleate cells. Each one of these cells contains one nucleus of each mating type and is the start of a new fruiting body

Binucleate cell multiplies by mitosis. Each of the binucleate cells that are formed will give rise to an ascus-initial cell when their nuclei fuse within the cell, so effecting fertilisation

Ascospores are released and germinate to give mycelium of two mating types, called '+' and '−'

HAPLOID

fertilisation

DIPLOID
$2n$ ascus
initial cell

ascus

mitosis   meiosis

Mature fruiting body containing numerous asci and ripe ascospores

Diploid nuclei in the ascus-initial cells undergo meiosis immediately

Each of the 4 haploid nuclei produced by meiosis then undergoes a mitosis to give an ascus containing 8 haploid ascospores (4 identical pairs)

Fig. 3.3 Highly simplified diagram of the life cycle of the fungus *Sordaria brevicollis*. *Sordaria* is an example of a haplontic organism which is haploid for most of its life cycle. The diploid phase is very brief and is restricted to the ascus-initial cells which undergo meiosis almost immediately after they are formed by fertilisation. The gametes are produced by mitosis and are in the form of pairs of nuclei which occur in binucleate cells in developing fruiting bodies.

In some groups of plants, such as the liverworts and mosses, the gametophyte is the dominant phase and comprises the main part of the plant body: the sporophyte is a small spore-producing structure which develops upon the gametophyte. In other groups the sporophyte is dominant, as in flowering plants, and the gametophytes are contained within the reproductive organs of the flower as tiny structures only a few cells in size. The ferns are intermediate. Their gametophyte and sporophyte phases are both well developed and grow as separate independent organisms. The trend in evolution has been towards a progressive reduction of the gametophyte and increasing dominance of the diploid sporophyte phase. We will follow this trend by briefly describing the life cycles of (1) a liverwort and (2) a moss, in which the gametophyte is predominant, (3) a fern, which has both phases well developed and (4) a flowering plant (angiosperm) in which the gametophyte shows the extreme form of reduction.

1. The liverwort, *Pellia epiphylla*   The predominant vegetative phase in *Pellia* is the haploid gametophyte. It consists of a simple **thallus** structure without any differentiation into stems or leaves (Fig. 3.4). The thallus produces the male and female sex organs, i.e. the **antheridia** and **archegonia**, which develop on its upper surface. When the antheridia rupture, the motile **spermatozoids** move in the film of water (from rain or dew), by action of their cilia, and fertilise the ova in the archegonia. The zygotes which are formed grow by mitosis to give the diploid spore-producing sporophytes. Haploid spores arise within the sporogonium by meiosis and germinate to give new thalli (Fig. 3.4).

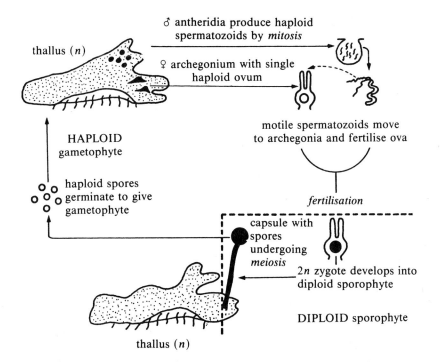

Fig. 3.4   Life cycle of the liverwort, *Pellia epiphylla*.

2. The moss, *Funaria hygrometrica*   The outline of the life cycle of a liverwort could just as easily be used for that of a moss such as *Funaria*. The major difference is in the structure of the haploid gametophyte.. In *Funaria* the gametophyte is the *leafy* moss plant which has an erect form with a 'stem' and numerous 'leaves', similar in appearance to those of the sporophyte of a higher plant. The male sex organs, antheridia, are formed at the apex of the main stem; and the female ones, archegonia, at the tip of the side branch. Fertilisation takes place in water when spermatozoids from ruptured antheridia are conveyed to the female gamete (ovum) in the archegonium, by rain or dew. As the diploid sporophyte develops from the zygote, the branch on which it is growing comes to form the main stem, and the male branch is pushed to one side as in Fig. 3.5.

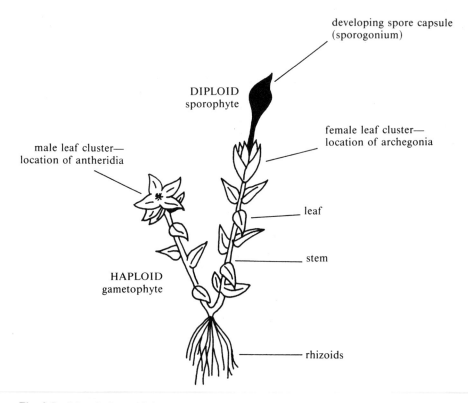

developing spore capsule (sporogonium)

DIPLOID sporophyte

female leaf cluster— location of archegonia

male leaf cluster— location of antheridia

leaf

stem

HAPLOID gametophyte

rhizoids

Fig. 3.5   Morphology of the moss, *Funaria hygrometrica.*

3. The fern, *Dryopteris*   In the fern the diploid sporophyte is the predominant phase of the life cycle. It is a large plant with compound leaves, or fronds, and a rhizome. It produces numerous haploid spores by meiosis. The gametophyte is a small haploid **prothallus** which arises from a germinating spore and grows as a completely separate and independent organism. Antheridia and archegonia are produced on the underside of the prothallus and the male spermatozoids and female ova unite in water in the archegonium. The diploid zygote grows initially on the prothallus and then develops into a new adult fern (Fig. 3.6).

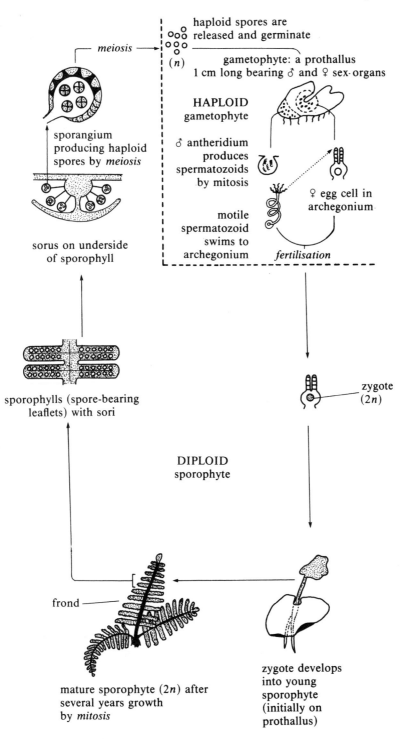

Fig. 3.6 Life cycle of the fern, *Dryopteris*.

4. The angiosperm, *Secale cereale* (rye)   In flowering plants the gametophyte is reduced down to a few cells. The diploid sporophyte is the plant itself which produces haploid male **microspores** and female **megaspores** by meiosis. Microspores are formed in the **anthers** and give rise to the **male gametophytes** or **pollen grains;** whilst megaspores are formed in the **ovaries** and become the **female gametophytes** or **embryo sacs**. A representative life cycle is that of the cereal rye, *Secale cereale*, shown in Fig. 3.7.

   (*a*) *Male gametophyte development.* The anthers contain numerous diploid **microspore mother cells,** or **pollen mother cells,** each of which undergoes meiosis to produce four haploid microspores. After separating from one another each microspore is encapsulated in a thick wall and becomes a pollen grain. The single haploid nucleus within the pollen grain divides by mitosis to give a **generative nucleus** and a **vegetative nucleus**. The generative nucleus divides again and produces two **male gametic nuclei** (Fig. 3.7).

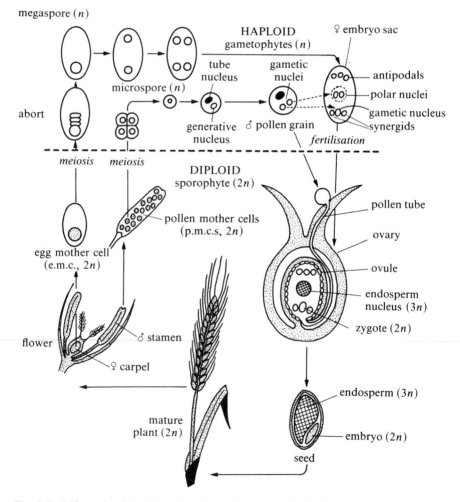

Fig. 3.7   Life cycle of the flowering plant *Secale cereale* (rye).

(b) *Female gametophyte development.* A single **egg mother cell** in each ovule undergoes meiosis to produce four haploid cells. Three of these abort and only one remains as a functional megaspore. The megaspore nucleus undergoes three successive mitotic divisions to give eight haploid nuclei in a special arrangement within the embryo sac (Fig. 3.7). When fully developed, the embryo sac is an 8-cell gametophyte without cell walls, corresponding to the fern prothallus (but with female sex cells only). The pollen is a 3-cell male gametophyte.

The spores which produce the gametophytes are not released from the parent (sporophyte) as they are in ferns, and the gametophytes do not grow independently of the parent but are retained in the 'body' of the sporophyte.

(c) *Pollination and fertilisation.* In flowering plants, male and female gametes are brought into contact after transfer of the pollen from the anther of one plant to the stigma of another. The transfer of pollen is principally by wind, as in rye, or by insects. After **pollination** has taken place the pollen grain germinates on the surface of the receptive stigma and produces a **pollen tube** which grows down between the cells of the style until it reaches the ovule (Fig. 3.7). The two male gametic nuclei then pass down the pollen tube, into the ovule, and **double fertilisation** takes place. One male nucleus fertilises the female gametic nucleus, to form the diploid zygote, and the other one combines with the two polar nuclei to form the triploid ($3n$) endosperm nucleus. The diploid zygote develops into the embryo and the endosperm nucleus gives rise to the endosperm storage tissue of the developing seed. When the seed is mature, it germinates and grows by mitosis into another sporophyte.

Diplontic cycles

In **diplontic** life cycles *the organism is entirely diploid except for the gametes.* The diploid adult produces haploid gametes directly by meiosis. These unite to form diploid zygotes which develop into new adults (Fig. 3.8).

This kind of life cycle is found universally among animals and is also known in some algae and some fungi. We will consider man as a representative and important example.

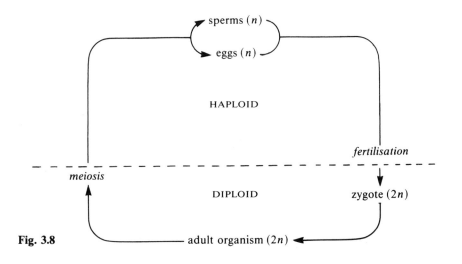

Fig. 3.8

In man, and in most other animals, the sexes are separate. Male and female diploid organisms grow and develop into adults by mitosis. Meiosis only occurs in the reproductive organs which are the **testes** in the male and **ovaries** in the female. In both sexes the haploid phase is restricted to a single-celled gamete, the **sperm** in the male and the egg, or **ovum**, in the female. Fertilisation of the ovum by a single sperm produces the diploid zygote which develops into a new individual. Our interest, in terms of chromosome cycles, is in seeing how meiosis is involved in the production of the male and female gametes—by **gametogenesis** (Fig. 3.9).

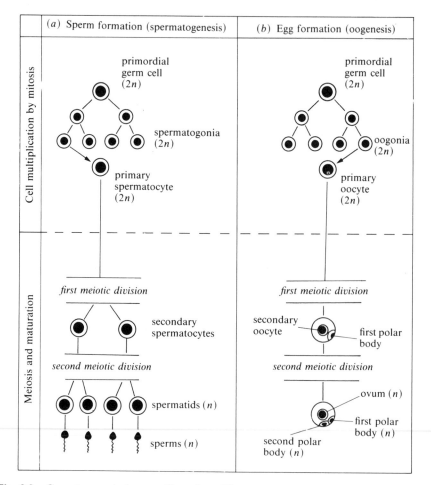

Fig. 3.9   Gametogenesis in man ($2n = 2x = 46$).

Spermatogenesis   In the testes **primordial germ cells** are multiplied up by mitosis to give large numbers of diploid **spermatogonia**. These enlarge and become the **primary spermatocytes**. The first meiotic division of the primary spermatocyte produces two **secondary spermatocytes.** The second meiotic division then produces four haploid **spermatids**. These mature into motile **spermatozoa (sperm)** which are transferred to the female during mating.

Oogenesis   Primordial germ cells in the ovary produce groups of diploid cells by mitosis. One cell in each of the groups becomes the **oogonium**, which enlarges into the **primary oocyte**. The first meiotic division in the primary oocyte gives one large cell, called the **secondary oocyte**, and a smaller one called the **first polar body**. Only the large secondary oocyte goes into the second division of meiosis. It forms a large haploid cell, the ovum, and a smaller cell which is the **second polar body**. The ovum is the functional female gamete. The polar bodies remain near the ovum but they are not involved in the union with the sperm.

Summary

The chapter describes the way in which mitosis and meiosis are involved in the three kinds of life cycles found in eukaryotic organisms to provide for their development and reproduction. In this book we are mainly concerned with the genetics of flowering plants and higher animals. Flowering plants are haplo-diplontic. They are diploid throughout the major part of their life cycle and have a very short haploid phase which is confined to the development of their pollen grains and embryo sacs. Animals are diplontic and produce haploid gametes directly by meiosis.

Questions on Chapters 2 and 3

1 Write an essay on 'chromosomes'.
2 By means of diagrams, describe the stages of cell division as it occurs at the root or shoot apex of a flowering plant.   *(S.U.J.B. Biol., 1983)*
3 List four distinct differences between the processes of meiosis and mitosis and state briefly the significance of each.
4 Discuss critically, with the aid of diagrams, the contention that meiosis is a reduction division followed by mitosis.   *(O.L.E. Bot., 1981)*
5 (*a*) Describe briefly the behaviour of the chromosomes during the process of mitosis.
   (*b*) List the places where you would expect mitosis to occur in a fully-grown higher plant.
   (*c*) Discuss the importance of mitosis in a fully-grown mammal.   *(C. Biol., 1981)*
6 For a cell with two pairs of chromosomes draw diagrams to show:
   (*a*) metaphase of mitosis
   (*b*) metaphase I of meiosis
   (*c*) prophase II of meiosis
   (*d*) metaphase II of meiosis.
7 (*a*) Give an illustrated account of meiosis in a plant cell. Indicate precisely what happens at the chiasmata.
   (*b*) What is the significance of meiosis?
   (*c*) State exactly in which cells it occurs and where these cells are situated in a flowering plant, a fern and an ascomycete fungus.   *(C. Bot., 1982)*
8 By means of a set of diagrams, illustrate the process of meiosis. Describe how this differs from the process of mitosis and explain the biological value of each.
   Which of the two processes operates in (*a*) the first and (*b*) the second division of the spermatocyte of a typical animal?   *(C. Zool., 1983)*
9 (*a*) Describe with the aid of diagrams the process of meiotic division.
   (*b*) Where, and at what stage in the life cycle, does meiosis occur in (i) a bryophyte, (ii) an angiosperm, (iii) a mammal?
   (*c*) What is the biological significance of meiosis?   *(C. Biol., 1983)*

10 (a) Compare the prophase of mitosis with the prophase of the first division of meiosis.

(b) Comment on the significance of any differences between these two types of prophase.

(c) Where does meiosis occur in the following: (i) a fern, (ii) a mammal, (iii) a flowering plant? *(L. Biol., 1979)*

11 (a) What is a chromosome?

(b) Describe what happens to a pair of homologous chromosomes during mitosis.

(c) Where may mitosis occur in (i) a mammal and (ii) a flowering plant? *(L. Biol., 1979)*

12 How are haploid pollen grains formed within the pollen sac of an angiosperm? Describe the germination of a pollen grain and its growth, including the process of fertilisation. *(C. Bot., 1981)*

13 (a) What is alternation of generations?

(b) Write an illustrated account of alternation of generations in the life cycle of a moss.

(c) How does alternation of generations in a fern and a flowering plant differ from that in a moss? *(L. Biol., 1979)*

14 (a) What is the significance of sexual reproduction?

(b) Describe how the gametes are brought together in (i) a mammal, (ii) a fern, (iii) a flowering plant. *(L. Biol., 1981)*

15 Write a comparative account of gametogenesis and fertilisation in the flowering plant and the mammal. *(W.J.E.C. Biol., 1979)*

16 Compare the alternation of haploid and diploid phases in the life cycles of selected examples of plants you have studied. What is its significance? What sorts of parallels exist in the animal kingdom? *(O.L.E. Biol., 1982)*

17 (a) State briefly the difference between asexual and sexual reproduction.

(b) Give an account of asexual reproduction in a named organism.

(c) What types of sexual cycles are found amongst plants and animals? Illustrate each type with a brief account of the life cycle of one named organism.

4
Mendelian inheritance I: Segregation

Observing chromosomes under the microscope as they undergo their meiotic reduction division is one way of studying the transmission of genetic material. Another way is to cross together two varieties of a species which differ in some obvious character, and then to study the pattern by which the two forms of the character are inherited from one generation to the next. This kind of breeding experiment was first used successfully by Gregor Mendel in the last century to unravel the laws of heredity in garden peas. It is a less direct method of experimentation than chromosome cytology because it involves making deductions about abstract 'genetic elements' which control characters, rather than looking directly at the genetic material itself.

In Chapters 4 and 5 we will deal with Mendel's breeding experiments and see how this work led to the discovery of the laws of heredity and to the concept that it is not the character itself which is transmitted during reproduction but some cell element, or factor, which determines the character. This factor we now call the **gene**.

Design of Mendel's experiments

Mendel set out to discover what happened to two alternative forms of a character when they were combined together to form a hybrid. To do this he took two varieties of the garden pea which differed in a single character (e.g. tall and dwarf for the character 'length of the stem'), crossed them together to make the hybrid, and then observed how the character was passed on over several generations of self-pollination. In this way he was able to find out about the nature of inheritance and the rules by which the transmission of a character could be predicted.

But how was it that Mendel was successful in this objective when others who had tried to do the same before him had failed?

It seems that he succeeded because he was an able and careful experimenter. He had a clear understanding of his objectives and of the way in which his experiments should be *designed* in order that he could interpret their outcome.

Choice of organism The pea, *Pisum sativum*, is an ideal choice of organism for experiments in heredity. There are different types which differ in discrete and easily recognisable characters. The pea is self-pollinating, the anthers and stigma being enclosed within a keel and naturally protected from the pollen of other pea plants (Fig. 4.1). Controlled crosses are easily made by opening

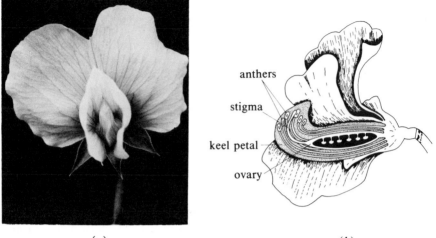

(a)   *(b)*

Fig. 4.1 (*a*) Photograph of the flower of the pea, *Pisum sativum*, at the time of pollination.
—(*b*) Drawing of side-view of the flower, with one side of the keel petal removed to show the way in which the anthers and stigma are enclosed within the keel resulting in self-pollination.

the keel petal, removing the anthers before they shed any pollen, dusting the stigma with pollen from the chosen male parent and then closing the keel for protection. The pea is easily grown and produces an abundance of progeny.

Determination of ratios   Mendel understood that it was necessary not only to cross together clearly distinct types, but also to count large numbers of offspring over several generations and to establish the **ratios** of their different forms. Earlier workers had made crosses between varieties of peas with contrasting pairs of characters and had noted the *mixture* of the two forms appearing in the later generations—but they had missed the vital observation which Mendel made that these mixtures actually contained the two types in **definite proportions** to one another.

Choice of characters   He confined his studies to seven carefully chosen characters, each of which was represented by discrete alternative pairs (e.g., tall and dwarf) without any intermediate forms (Box 4.1, Figs 4.4 and 5.1).

Verification of pure line   Before using any of his pea varieties in crossing experiments, Mendel first established that the seven pairs of alternative characters remained constant when his lines were allowed to self-pollinate naturally for several generations. In other words he confirmed that for any given character form he had a **pure line** which was **pure-breeding** (or **true-breeding**) for that character: seeds from tall plants, for example, always gave rise to only tall plants.

Monohybrid crosses   He crossed together plants which differed in only *one* character, e.g. round × wrinkled seeds (Fig. 5.1). We now refer to this as a

monohybrid cross. He then followed the inheritance of round and wrinkled through several generations of self-pollination ('selfing'). In working with this single character, seed shape, he simply ignored the presence of the other six characters. These he tested separately and so made seven different monohybrid crosses, for the seven pairs of characters, and studied the inheritance of each pair of them individually. His predecessors had often been confused because they had tried to study the inheritance of several pairs of characters all at the same time.

Controls   The other sound measure which he took was to grow uncrossed pure-breeding plants, under the same conditions, to serve as controls when evaluating (or sorting out the characters of) the progeny of the various hybrids.

Monohybrid crosses

The seven characters behaved in the same way in all of the monohybrid crosses so we may consider just one of them—length of the stem—as an example and make reference to the other six as appropriate.

The F_1 hybrids

Crosses were made so that each form of the character was used as both male (pollen donor) and female (egg donor) parent. For example, pollen from tall plants was used to fertilise short plants and vice versa. This is called a **reciprocal cross**:

$$\left.\begin{array}{l} \text{♀ tall} \times \text{dwarf ♂} \\ \text{♀ dwarf} \times \text{tall ♂} \end{array}\right\} \text{reciprocal cross}$$

Hybrids from these crosses we now call the **first filial**, or F_1, generation and Mendel found that when the seeds were grown up the F_1 were all identical to one another with respect to height: they all resembled the tall parent and there were no short or intermediate forms. Both the tall and short characters must be present in the hybrid but only the tall character was apparent or **expressed**. He named this the **dominant** form. The dwarf character which did not appear he termed **recessive**. He also found that tall was dominant regardless of whether it had been contributed by the female or the male parent, and that the *direction* of the cross was therefore immaterial (Fig. 4.2).

Parents	♀ tall × dwarf ♂	♀ dwarf × tall ♂
	↓	↓
F_1	all tall	all tall

Fig. 4.2   (♀ = female; ♂ = male).

The F_2 generation

The **second filial** generation, or F_2, was raised by simply allowing the F_1 hybrids to self-pollinate. In the F_2 the recessive character reappeared together with the dominant one. Furthermore when the numbers were counted it turned out that the dominant and recessive forms were in an approximate ratio of 3 tall : 1

dwarf. The same result was found for all six characters. A summary of Mendel's original data from all of his monohybrid crosses is given in Box 4.1.

Box 4.1. Inheritance in the pea

The following is a summary of Mendel's results on inheritance in the pea, showing the $3:1$ ratio which he obtained in all seven of his monohybrid crosses. The large F_2 sample in each of the crosses is a total of several F_2 families produced from a number of F_1 plants.

Character	Cross	No. of F_2 counted	No. showing dominant character	No. showing recessive character	Ratio
Shape of seed	round × wrinkled	7324	5474 round	1850 wrinkled	2.96:1
Colour of cotyledons	yellow × green	8023	6022 yellow	2001 green	3.01:1
Colour of seed coat	grey × white	929	705 grey	224 white	3.15:1
Shape of pods	inflated × constricted	1181	882 inflated	299 constricted	2.95:1
Colour of pods	green × yellow	580	428 green	152 yellow	2.82:1
Position of flowers	axial × terminal	858	651 axial	207 terminal	3.14:1
Length of stem	tall × dwarf	1064	787 tall	277 short	2.84:1

The F_3 generation

The breeding behaviour of the F_2 was established by allowing them to self-pollinate and then examining a sample of their progeny. Those which showed the recessive character were found to be pure-breeding and to give all dwarf progeny in the **third filial** generation, the F_3. One in three of the tall dominants were also pure-breeding and gave all tall F_3. The other two thirds were classified as 'impure' **mixed-breeding** dominants and yielded F_3 families with a $3:1$ ratio of dominant to recessive types. This breeding pattern, from the parents down to the F_3 generation, was found to be the same for all seven characters, and may be summarised in terms of tall and dwarf as shown in Fig. 4.3.

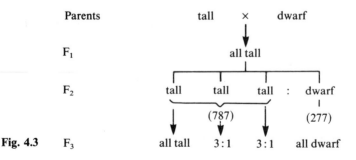

Fig. 4.3

From studying these F_3 progeny Mendel concluded that the $3:1$ ratio observed in the F_2 could be further resolved into a $1:2:1$ ratio:

$$1 \quad : \quad 2 \quad : \quad 1$$

pure-breeding dominants   mixed-breeding dominants   pure-breeding recessive

Fig. 4.4   Two-week old pea seedlings showing alternative forms of the character 'length of the stem'.

Interpretation of the monohybrid crosses

The spark of genius which Mendel showed in interpreting this pattern of inheritance was in seeing that the $1:2:1$ ratio into which the F_2 could be resolved represents the product of the binomial expression derived from randomly combining two pairs of unlike 'elements'.

$$(A+a)(A+a) = 1AA + 2Aa + 1aa$$

The **Punnett Square method**, which was later devised by R. C. Punnett, shows us more clearly how this combining process takes place (Fig. 4.5).

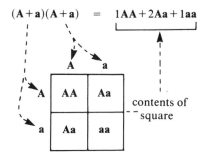

Fig. 4.5   Punnett Square.

It was the realisation of the way in which the $1:2:1$ ratio in the F_2 could arise, which gave Mendel the idea that it is not the character itself which is passed on during reproduction, but rather some kind of particulate factor which determines the character and which exists in the cells in two dissimilar forms $(A+a)$. The factors must occur singly in the gametes and are able to combine together randomly in pairs according to the binomial formula to give the three F_2 breeding types in the proportions of $1:2:1$. With this idea in mind he was able to interpret the whole of his monohybrid experiment and to arrive at an understanding of the basic mechanism of heredity. The reasoning behind the interpretation, in terms of the cross tall × dwarf, is summarised as follows:

1. As both the tall and the dwarf characters reappear in the F_2 there must have been a contribution to the F_1 hybrid by both parents. The F_1 received something, a 'cell-element' or 'factor' which determines tallness from the tall parent and a factor which determines dwarfness from the dwarf parent. *These cell elements, or factors, we now call* **genes**.
2. Since both forms of the cell element are present in the F_1 the character must be controlled by a pair of elements: one dominant and one recessive. *The alternative forms of the cell elements are now called* **alleles**.
3. It follows that the cell elements exist in pairs in the cells of the plants, and that in the formation of the gametes only one member of each pair is transmitted into each gamete.
4. In forming the hybrid the two different elements are brought together, without any mixing, or blending, between them, and when the gametes are produced by the F_1 they separate quite cleanly from one another to give two different kinds of gametes. The process by which this separation takes place is now called **segregation**.
5. The F_2 ratio results from random combinations of the two kinds of the female and male gametes at fertilisation.

To explain how the F_2 ratios arise, the interpretation is shown in the form of a diagram in Fig. 4.6. In this diagram we have used symbols to represent the two forms of the alleles: The bold capital letter **T** denotes the dominant allele and the bold lower-case letter **t** the recessive. The parents are pure-breeding and produce only one kind of gamete and they must therefore contain an identical pair of alleles. We can represent them as **TT** for pure-breeding tall and **tt** for pure-breeding dwarf: *individuals which carry two identical alleles of a gene are now said to be* **homozygous**. The F_1 hybrids received one allele from each parent and therefore contain an unlike pair, **Tt** (i.e. they are **heterozygous**). They appear tall because **T** is dominant over **t**.

We can see from this method of representation in Fig. 4.6 how it is that the $1:2:1$ ratio is the result of the chance, or random, combinations of the two kinds of male and female gametes at fertilisation. On the male side we have $\frac{1}{2}$**T** and $\frac{1}{2}$**t** in a sample of gametes. If the half that are **T** have an equal chance of combining with egg cells that are either **T** or **t** from the female side, then $\frac{1}{2}$ of a $\frac{1}{2}$, or $\frac{1}{4}$, will be **TT** and $\frac{1}{2} \times \frac{1}{2}$, or $\frac{1}{4}$, will be **Tt**. Likewise the half of the male gametes that are **t** have an equal chance of fusing with **T** or **t** eggs so that we have a further $\frac{1}{2} \times \frac{1}{2}$, or $\frac{1}{4}$**Tt**, and $\frac{1}{2} \times \frac{1}{2}$, or $\frac{1}{4}$**tt**.

In observing the F_2, of course, all we see from the appearance of the plants is the ratio of 3 tall : 1 dwarf. Their actual genetic constitution was determined by allowing them to self-pollinate and by seeing what turned up in a sample of the F_3 progeny. In this way the $1:2:1$ ratio was revealed because the dwarfs yielded all dwarf progeny, confirming that they were homozygous recessive (**tt**); one third of the talls gave all tall F_3 families and must therefore have been pure-breeding homozygous dominant (**TT**), and the other two thirds segregated again to give 3 tall : 1 dwarf in their F_3 families, thereby confirming their mixed-breeding heterozygous status as **Tt**.

From what has been said above it is obvious that two plants which look identical can be of different genetic constitution, because of the dominance

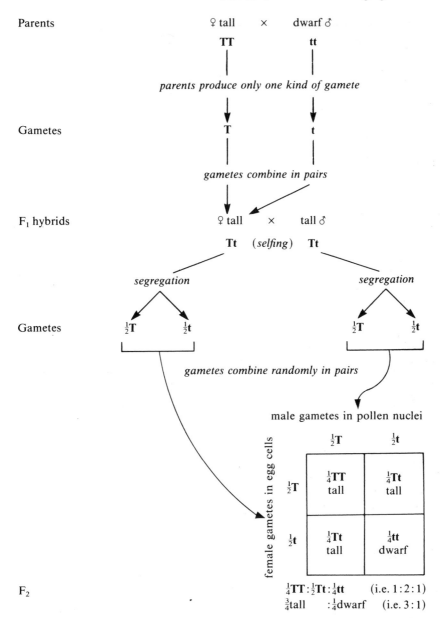

Fig. 4.6 Interpretation of Mendel's monohybrid cross between parent pea plants which differ in a single character. In this example, the character is length of the stem and the alternative forms are tall and dwarf plants. The pattern of inheritance is explained by postulating the existence of a pair of dissimilar factors in the cells of the parents which are passed on singly through the gametes and combine together again in the F_1. One of the factors is dominant over the other one which explains why all the F_1 resemble the tall parent. The factors segregate from one another during sex cell formation in the F_1 to give two kinds of gametes in equal frequencies. Random combination of the male and female gametes explains the F_2 ratio of 3 tall:1 dwarf. (In the F_1 the cross ♀tall×tall♂ refers to female and male parts of the same flower.)

of one allele over another. *The genetic constitution of an individual,* which is determined by crossing experiments, is called the **genotype**. The appearance of an organism is its phenotype. Both **TT** and **Tt** are identical in phenotype (tall) but they behave differently in genetic crosses, because they are of different genotype. The environment can also affect the appearance of an organism, as was explained in Chapter 1. The **phenotype** is therefore defined as *the appearance and function of an organism as a result of its genotype and its environment.*

The law of segregation

This interpretation of the result of the monohybrid crosses gave Mendel his basic theory—he saw that inheritance could only be explained by the existence of character elements in *pairs*; by their separation *singly* into gametes in equal frequencies; and by their random combinations back into pairs when the gametes fuse to form the zygotes of the next generation.

Contrary to popular belief, however, Mendel never presented the conclusions of his work in the form of the two laws that are generally attributed to him. It was his successors who redefined his work and made it into the two laws with which we are now familiar. His experiments on the monohybrid crosses thus became known as 'Mendel's First Law' or 'The Law of Segregation'. Because different workers interpreted his work in their own ways there is no *definitive* version of this law. When we speak of the original form of Mendel's Laws, what we are referring to is the redefinition of his work, as others saw it, in terms of his breeding experiments. Put in these original terms 'Mendel's First Law' or **The Law of Segregation**, may be stated as follows: *the characters of an organism are determined by pairs of factors of which only one can be present in each gamete.*

Likewise there are numerous modern definitions of this law. These take account of the introduction of new terms and make references to the discovery of the relation between genes and chromosomes, which is discussed in Chapter 6. Stated in modern terms **The Law of Segregation** says that: *contrasting forms of a character are controlled by pairs of unlike alleles which separate in equal numbers into the different gametes as a result of meiosis.*

The testcross

To confirm the assumption . . . 'that the various (two) kinds of egg and pollen cells were formed in the hybrids on the average in equal numbers' . . . Mendel later devised a simple breeding experiment, i.e. a testcross, in which his hybrids (**Tt**) were crossed to a pure-breeding recessive homozygote (**tt**). Since the recessive homozygote produces only one kind of gamete, the expectation is that the testcross progeny will be of two kinds in equal frequencies (Fig. 4.7).

The numbers observed were 87 tall (**Tt**) and 79 dwarf (**tt**), which Mendel took to confirm the assumption.

This simple breeding test, in which *a heterozygote is crossed with the corresponding recessive homozygote* in order to verify the genetic constitution of the heterozygote, is now a standard method of genetic analysis and is known as a **testcross**. In the particular case where *an individual is crossed with one of its parents,* for the same purpose, it is referred to as a **backcross**.

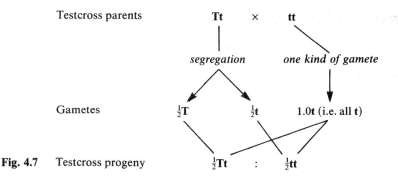

Testcross parents   Tt   ×   tt

segregation   *one kind of gamete*

Gametes $\frac{1}{2}$T $\frac{1}{2}$t   1.0t (i.e. all t)

Fig. 4.7   Testcross progeny $\frac{1}{2}$Tt   : $\frac{1}{2}$tt

Box 4.2 Sampling error and probability

In the monohybrid cross AA × aa the law, or principle, of segregation tells us that the two kinds of F_1 gametes will be produced in equal numbers. At fertilisation they will combine together randomly in pairs to give a ratio of 3 : 1 of the dominant and recessive phenotypes in the F_2. What we observe in an actual experiment, however, may not always agree exactly with our theoretical expectation. If we have only four F_2 progeny, for example, it is most unlikely that we will find the exact ratio of 3 : 1. The reason is that chance is involved in the production of the two classes of dissimilar gametes from the heterozygotes as well as in the way in which they come together at fertilisation.

If we were to shake out a small sample of pollen grains from an anther of a plant which is heterozygous (Aa), the observed numbers of the two classes of haploid grains, A and a, may deviate quite widely from the expected 1:1 by chance alone. A sample of only six grains may well contain the unrepresentative numbers of 4 A + 2 a. Another small sample, by chance, may contain 2 A + 4 a. In a much larger sample these small chance deviations will cancel one another out, and the frequencies we find will be much more representative of the actual population of gametes. On the female side, in flowering plants and in animals, only one of the products of meiosis functions as the egg and it is a matter of chance whether it carries the dominant or recessive allele.

At fertilisation the random combinations of sperms and eggs to give zygote proportions of 1 AA : 2 Aa : 1 aa are subject to the same chance effect as we have in tossing two coins and testing for the head/tail combinations 1 HH : 2 HT : 1 TT. In other words our F_2 ratio is subject to **sampling error**, and the smaller the number of progeny we are dealing with the greater the error. For this reason, genetic ratios have to be tested by statistical analysis to see whether the deviation of the observed and expected values is acceptable on the basis of chance, or whether there is some real deviation, i.e. the data do not fit the expected ratio. These statistical procedures are dealt with in Box 5.3. Mendel was well aware of the problem of sampling error and this is the reason why he worked with such large numbers of F_2s.

We can deal with this element of chance by expressing our expectations for a particular cross in terms of probabilities. We say that for the monohybrid cross the probability is 1 in 4 that an individual F_2 will have the recessive phenotype and 3 in 4 that it will be dominant. In this way we can say what our expectations are even when we have only one offspring. For an *individual* of course these expectations are the same regardless of the sample size. The other way of expressing the probability is to say that we expect three quarters of the sample to be of dominant phenotype and one quarter recessive: in this case, though, our chances of fulfilling the expectation will be better when the sample is large.

Examples of Mendelian inheritance in other species

Following the rediscovery of Mendel's work at the turn of the century, it was quickly found that the basic principles of heredity which he had discovered in peas applied to numerous other species of plants and animals as well.

In this section we will consider some examples of Mendelian heredity in species other than the pea, and see how we can use the principle of segregation in solving some practical problems.

Maize

Maize (*Zea mays*), or corn as it is called in the U.S.A., is a favoured organism for studies on genetics. The reason is clear to see in the example shown in Fig. 4.8. The ear, or cob, can have 500 or more individual kernels on it and there are numerous different forms available affecting colour and shape. The bulk of the kernel (= seed) is made up of the endosperm and embryo, both of which are structures resulting from the fertilised egg, and any segregating character affecting these tissues can be classified from the appearance of the mature seed when the cob is harvested.

Fig. 4.8   An ear of maize (corn) segregating in a 1:1 ratio of purple:yellow for the character seed colour. This example is from a testcross of heterozygous purple (**Pp**) × pure-breeding homozygous yellow (**pp**).

The ear shown in Fig. 4.8 is from the cross of a plant grown from a purple seed with one known to be a pure-breeding recessive homozygote for yellow seeds (**pp**), and as we can see it has a mixture of purple and yellow seeds (the colour is due to the endosperm). On counting this cob it turned out that there were 254 purple and 234 yellow seeds, which we recognise as approximating to a 1:1 ratio. From this sample of 488 the expected numbers for a 1:1 ratio are 244:244, so we see that the deviations of +10 and −10 are quite small—in fact they are within the acceptable limits for sampling error. Since there are only two classes, we can conclude that we are dealing with a single character with two forms, and that the purple plant is a heterozygote segregating for two alleles (Fig. 4.9).

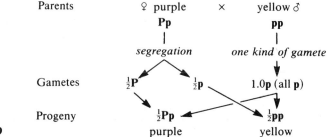

Fig. 4.9

Leopard

The black panther is simply a black variety of the spotted leopard (*Panthera pardus*). It occurs quite commonly in areas of tropical or semi-tropical rain forests, where it appears to compete successfully with the normal spotted form. Although we don't know for certain, a likely explanation for its success in the rain forests is that the black colour provides an excellent camouflage in the gloom of the dense forests and this gives it the advantage of concealment when hunting its prey. On the open plains the black panther is much rarer, presumably because it is much more conspicuous against the open grassland, whilst the spotted coat of the normal form blends in perfectly.

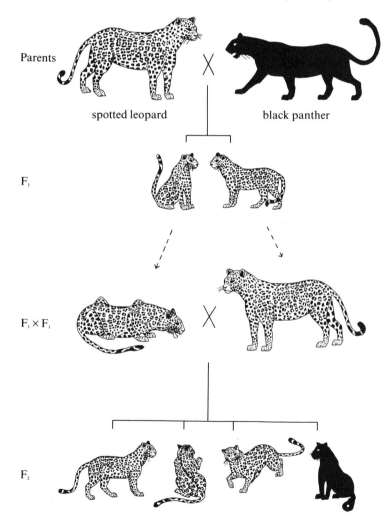

Fig. 4.10   The spotted leopard and the black panther are both varieties of the species *Panthera pardus*. Crosses between the two forms in captivity have shown that the coat-colour character is determined by a single gene with two alleles. This is known because spotted is dominant over black in the F_1, and statistics compiled from breeding pairs in several zoos show a 3:1 ratio of spotted:black in the F_2.

In zoos the black and spotted forms interbreed freely and statistics have been collected from a large number of different zoos in order to establish the hereditary basis of the variation. As Fig. 4.10 shows, the outcome is quite straightforward. The 3 : 1 ratio observed in the F_2 tells us immediately that we are dealing with a monohybrid cross and that the character is determined by a single gene. The F_1 cats are all spotted, so we can say that this is the dominant form of the character. With this knowledge we can interpret the breeding pattern (Fig. 4.11).

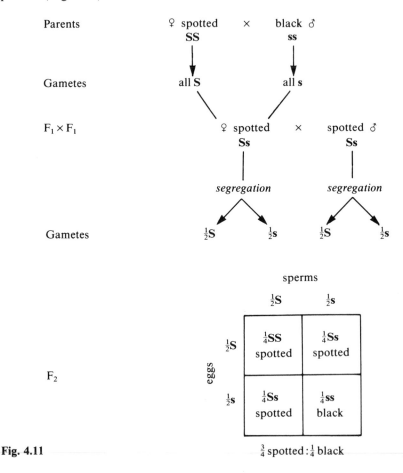

Fig. 4.11 $\frac{3}{4}$ spotted : $\frac{1}{4}$ black

Man

There are a number of familiar and distinctive characters in man (*Homo sapiens*) which are due to a single gene difference and which show typical Mendelian monohybrid inheritance. Experimental crosses are not possible, of course, but information about heredity can be pooled from different families and from patterns of transmission of characters traced through pedigrees of ancestors and descendants. Once the genetic control of a character has been worked out we can use our knowledge of Mendelian principles to make predictions about the outcome of certain marriages, and about the probabilities with which particular genotypes or phenotypes will occur among the children.

Albinism is a well known recessive mutation which is caused by a block in one of the chemical processes leading to production of the pigment melanin. The appearance of the phenotype of homozygotes includes the features of white hair, light-coloured skin and pink eyes. Because the mutation is recessive it is hidden in heterozygotes, i.e. **carriers**, who are capable of transmitting the gene to their progeny. The law of segregation enables us to interpret pedigrees and to make predictions about the kinds of children expected from matings between various genotypes. Some examples follow.

1. ♀ **Normal homozygote × albino** ♂

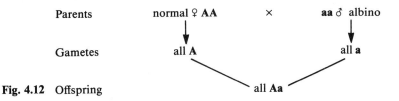

Fig. 4.12 Offspring

From this cross all the children will be carriers, and will have a normal phenotype. The probability of being normal is 1.0.

2. ♀ **Carrier × albino** ♂

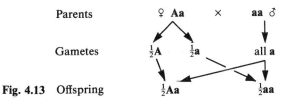

Fig. 4.13 Offspring

Half the children are expected to be normal and half albino, but in a small family this expectation may not be realised due to sampling error. A single child has a probability of $\frac{1}{2}$ of being normal and $\frac{1}{2}$ of being albino. If five albinos are born 'in a row' the sixth one still has a probability $\frac{1}{2}$ of being normal and $\frac{1}{2}$ of being albino—in the same way as when we toss a coin five times and obtain all heads the probability for the sixth toss is still $\frac{1}{2}$H + $\frac{1}{2}$T.

Box 4.3   Tree diagrams

Another question we may ask is: if a couple **Aa × aa** intend having only two children, what is the probability that both will be albino, both will be normal or they will be one of each?

The answer is worked out as follows: we say that the probabilities for the first one are $\frac{1}{2}$ normal + $\frac{1}{2}$ albino. Now for each of these possibilities the probabilities for the second child are $\frac{1}{2}$ normal + $\frac{1}{2}$ albino again (see Fig. 4.14).

The probability of having two normals or two albinos is $\frac{1}{4}$, the probability of having one of each is $\frac{1}{2}$, twice as great. Expectations for combinations with more than two children can be worked out in the same way by extending the tree diagram and multiplying through the probabilities.

1st child	2nd child	both children	
$\frac{1}{2}$ normal	$\frac{1}{2}$ normal →	normal + normal $= \frac{1}{2} \times \frac{1}{2} = \frac{1}{4}$	normal + albino $= \frac{1}{2}$
	$\frac{1}{2}$ albino →	normal + albino $= \frac{1}{2} \times \frac{1}{2} = \frac{1}{4}$	
$\frac{1}{2}$ albino	$\frac{1}{2}$ normal →	albino + normal $= \frac{1}{2} \times \frac{1}{2} = \frac{1}{4}$	
	$\frac{1}{2}$ albino →	albino + albino $= \frac{1}{2} \times \frac{1}{2} = \frac{1}{4}$	

Fig. 4.14

3. ♀ Normal phenotype × albino ♂

♀ AA/Aa? × aa ♂

If this couple only have 2 or 3 children and there is an albino among them, we can be certain that the mother is heterozygous. If they are normal can we be sure she is homozygous?

Where the pattern of inheritance of a certain character is unknown, its genetic basis can often be worked out for the first time by constructing a pedigree. A **pedigree** is simply *a diagram of a family tree over a number of generations showing how the ancestors and descendants are related to one another.* The sex of the individuals and their phenotype with respect to a given character can be represented using symbols.

The example shown in Fig. 4.15 is a pedigree for **brachydactyly** (very short fingers) and is one of the first studies of single gene inheritance in man. It is an excellent example of the principle of segregation in human genetics.

The initial marriage between an affected female and a normal male produced eight children in the next generation (labelled II): four were brachydactylous

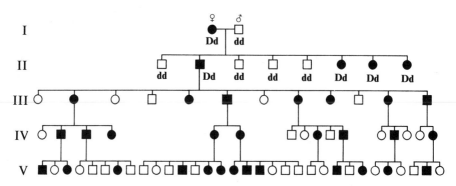

Fig. 4.15 Pedigree of the 'short finger' character, brachydactyly, in man. Females are represented by circles and males by squares. A black symbol indicates an individual displaying the character. From the pattern of segregation in this pedigree it can be deduced that brachydactyly is a dominant character determined by a single gene. The genotypes are shown for the parents and the next generation only. The spouses, except for the original parents, were all normal (**dd**) and they have been left out of the pedigree so that the 1:1 segregation pattern can be seen more clearly.

and four were normal. These numbers are reminiscent of a testcross between a heterozygote and a recessive homozygote, and they suggest that the mutation causing brachydactyly is dominant: $♀Dd \times dd♂$. (If it were recessive all the children would have been normal.) The suggestion is confirmed by the outcome of the subsequent generations in the pedigree. In each generation, approximately half of the children are brachydactylous and half are normal.

Summary

Mendel discovered that characters are determined by pairs of cell elements, now called genes, which are transmitted from one generation to the next in a regular and predictable manner. He made his discovery because of the careful way in which he designed his experiments and because his knowledge of mathematics enabled him to interpret his results after carefully counting the numbers of the two kinds of progeny which segregated out in a fixed $3:1$ ratio in the F_2. The results of the experiments on the monohybrid crosses were later formalised as Mendel's First Law, or the Law of Segregation.

Questions and problems follow Chapter 5.

5
Mendelian inheritance II: Independent segregation

Having resolved the issue of the monohybrid crosses the next thing which Mendel wanted to know was: what happens when *two* pairs of contrasting characters are combined together in the same hybrid? To answer this question he made *crosses between pure lines differing in two characters*—that is **dihybrid crosses**.

Dihybrid crosses

We will consider the experiment in which he followed the inheritance of the two characters 'form of the seed' and 'cotyledon colour'. It was already known from the monohybrid crosses that round seed form was dominant over wrinkled, and that yellow cotyledon colour was dominant over green. The round and wrinkled phenotypes are shown in Fig. 5.1.

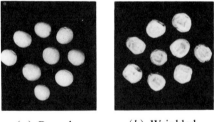

(*a*) Round   (*b*) Wrinkled

Fig. 5.1 Mature pea seeds showing alternative forms of the character 'seed shape'.

Mendel crossed together parent plants which had round seeds and yellow cotyledons with those having wrinkled seeds and green cotyledons. The F_1 all had the dominant characters of round yellow, and on selfing ($F_1 \times F_1$) they gave rise to an F_2 with four different phenotypes in a ratio of $9:3:3:1$. Two of the phenotypes resembled the parents and two of them were new types combining together characters from both of the parents (Fig. 5.2).

In interpreting this result Mendel made the important observation that when the characters of the F_2 phenotypes are looked at individually the pairs of alternative characters are in the ratio of 3 round:1 wrinkled and 3 yellow:1 green—the same as they were in the monohybrid crosses that he had studied previously. From this observation he saw immediately that the $9:3:3:1$ ratio arises simply because the two $3:1$ ratios are associated together in the same

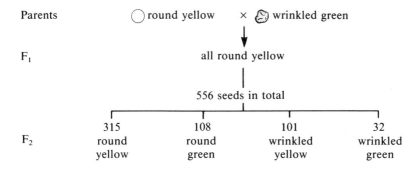

Parents   ◯ round yellow   × 🌰 wrinkled green

F₁   all round yellow

                    556 seeds in total

F₂   315   108   101   32
        round   round   wrinkled   wrinkled
        yellow   green   yellow   green

Fig. 5.2

cross and that the two pairs of characters behave quite independently of one another in their inheritance. The same theory of combining pairs of cell elements applies as before—but now there are two pairs being transmitted at the same time and this gives rise to four kinds of gametes in both the female and the male, and 4×4 ways in which they can combine together to give the F_2. The interpretation of the experiment is shown in Fig. 5.3 overleaf. The symbols **R** and **r** are used for the dominant and recessive alleles of the gene for seed form, and **Y** and **y** for the dominant and recessive alleles of the gene for cotyledon colour.

The parents are pure-breeding and have the homozygous combinations of alleles **RRYY** for round yellow and **rryy** for wrinkled green. They produce only one kind of gamete: **RRYY** parents produce gametes with **RY** only, and **rryy** parents produce only **ry** gametes. The F_1 are therefore all identical and heterozygous for both pairs of alleles (**RrYy**). Each gamete produced by the F_1 must contain one allele from each gene pair. When the dominant allele **R** passes into a gamete, it has an equal chance of being accompanied by either **Y** or **y**; likewise the recessive **r** also has equal chances of going with **Y** or **y**.

Since **R** and **r** also segregate in equal proportions, four kinds of gametes are produced in equal frequencies (Fig. 5.4.).

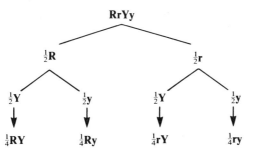

                        RrYy

            ½R   ½r

      ½Y   ½y   ½Y   ½y

Fig. 5.4   ¼RY   ¼Ry   ¼rY   ¼ry

The Punnett square method of combining these gametes shows us how the 16 F_2 combinations are formed and how the $9:3:3:1$ ratio of phenotypes arises. The essential point to note is that the F_2 ratio is dependent upon there being four kinds of gametes present in equal frequencies, and that the gametes are produced in this way because the two pairs of alleles segregate independently of one another during gamete formation.

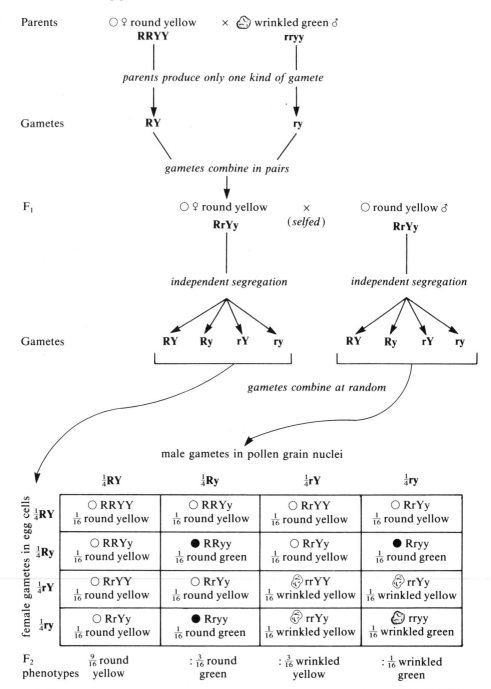

Fig. 5.3 Interpretation of the outcome of a dihybrid cross between parent pea plants which differ in two pairs of contrasting characters. Each character is controlled by a pair of alleles which segregate independently of one another from the heterozygous F_1. Random combinations of the four kinds of gametes gives rise to the $9:3:3:1$ ratio of phenotypes in the F_2 generation.

Confirming the ratio

All that we observe in the F_2 is the four phenotypic classes of pea seeds in the ratio of $9:3:3:1$. The existence of the nine different genotypes, and of the four kinds of egg cells and pollen nuclei in equal frequencies is 'assumed'—as Mendel put it—from the interpretation of the ratio. In order to verify (1) that the nine genotypes were present in their assumed proportions and (2) that four kinds of male and female gametes were produced in equal frequencies he carried out two further kinds of breeding tests, in the same way as had been done for the monohybrid crosses:

1. He grew all 556 of the F_2 seeds and then allowed the resulting plants to self-pollinate in order to determine their genotypes by looking at the F_3 progeny. The wrinkled green seeds were found to give plants which were pure-breeding with all wrinkled green seeds, confirming that they were homozygous for both pairs of alleles (**rryy**), as assumed. One third of wrinkled yellow seeds (i.e. 28 out of the 96 that grew) gave plants which were pure-breeding for both pairs of alleles confirming their genotype as **rrYY**, and the other two thirds (68/96) were pure-breeding for wrinkled and mixed-breeding for yellow, showing that they were **rrYy**. In the same way the selfing of round green and round yellow gave a mixture of pure-breeding and impure mixed-breeding types which were in the right proportions to conform with their assumed genotypes.

2. To demonstrate that the different types of gametes were produced by the hybrids in equal frequencies, some F_1 plants were testcrossed to the double-recessive parental type:

$$F_1 \text{ hybrid } \mathbf{RrYy} \times \mathbf{rryy} \text{ testcross parent}$$

The idea of this testcross is that the double-recessive genotype produces only one type of gamete which carries the two *recessive* alleles **ry**, and because of this the phenotype of the progeny will *show up* the genotypes of the F_1 gametes from which they were formed. If the assumption about the gametes is correct then the four phenotypes of the testcross progeny should be present in the proportions shown in Fig. 5.5.

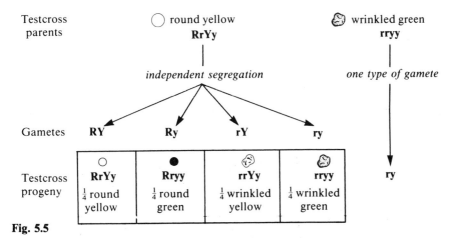

Fig. 5.5

This is exactly what Mendel found. In a sample of 208 progeny there were 55 round yellow, 51 round green, 49 wrinkled yellow and 53 wrinkled green. These observed values approximate very closely to the expected ones of $52:52:52:52$, for a sample of that size, and give direct confirmation that the four kinds of gametes are produced in their assumed frequencies.

The law of independent segregation

Mendel continued his studies by making experiments involving three pairs of contrasting characters. He found that the characters behaved in the same independent way and that all combined randomly with one another in the F_2. In general terms three pairs of alleles in a cross, **AABBCC** × **aabbcc**, give a **trihybrid** F_1 which produces eight different gametes and on selfing gives an F_2 with eight phenotypes in the ratio of $27:9:9:9:3:3:3:1$.

The experiments on the dihybrid and trihybrid crosses confirmed Mendel's theory. He saw that the way in which the unlike pairs of factors in a hybrid could separate from one another into the gametes and then come together again in predictable combinations to give the F_2, worked in the same way regardless of the number of different character pairs that were involved in the cross.

This part of his work later became known as 'Mendel's Second Law'—'The Law of Independent Assortment' or 'The Law of Independent Segregation'. There is no definitive version of the law, but defined in terms of the breeding experiment results alone it can be stated as follows: *when two or more pairs of characters are brought together in a cross they segregate independently of each other.* Expressed in present day terminology the 'law' is somewhat different and now takes account of the known chromosomal basis of heredity. **The Law of Independent Segregation** is the most widely used term and it may be defined thus: *two or more unlike pairs of alleles segregate independently of each other as a result of meiosis provided the genes concerned are unlinked.* In Chapter 6 we will explain the chromosomal basis of independent segregation and how it is that the law only applies to genes which are unlinked, i.e. located on different pairs of chromosomes.

Box 5.1   Sampling error and probability

The question of sampling error arises in dihybrid crosses just as it did in monohybrid crosses discussed in Box 4.2.

In a very small sample of F_2s it is meaningless to say that we expect to get the four phenotypic classes of progeny in the exact proportions of $\frac{9}{16}:\frac{3}{16}:\frac{3}{16}:\frac{1}{16}$. We may have less than 16 progeny—then what are our expectations? The answer is to use probabilities. We simply have to say that for an individual seed the probability that it will be green wrinkled is 1 in 16; that it will be yellow wrinkled 3 in 16, and so on. The probabilities can be obtained from the Punnett Square, or by simple multiplication of the probabilities for the separate characters:

The expectation
for round and wrinkled is $\frac{3}{4}$ round: $\frac{1}{4}$ wrinkled,
for yellow and green is $\frac{3}{4}$ yellow: $\frac{1}{4}$ green.

As the two characters are independent of one another, the probabilities for the various combinations of them are the product of their separate probabilities:

$$\text{round yellow} = \tfrac{3}{4} \times \tfrac{3}{4} = \tfrac{9}{16}$$
$$\text{round green} = \tfrac{3}{4} \times \tfrac{1}{4} = \tfrac{3}{16}$$
$$\text{yellow wrinkled} = \tfrac{3}{4} \times \tfrac{1}{4} = \tfrac{3}{16}$$
$$\text{green wrinkled} = \tfrac{1}{4} \times \tfrac{1}{4} = \tfrac{1}{16}$$

What is the probability of getting pure-breeding round green (**RRyy**)?

The probability of finding the **RR** genotype for the seed form character is 1 in 4, because in the F_2 we have:

$$\tfrac{1}{4}\textbf{RR} + \tfrac{1}{2}\textbf{Rr} + \tfrac{1}{4}\textbf{rr}$$

The probability of getting pure-breeding green (**yy**) is also 1 in 4:

$$\tfrac{1}{4}\textbf{YY} + \tfrac{1}{2}\textbf{Yy} + \tfrac{1}{4}\textbf{yy}$$

The probability of obtaining pure-breeding round green is therefore:

$$\tfrac{1}{4}\textbf{RR} \times \tfrac{1}{4}\textbf{yy} = \tfrac{1}{16}\textbf{RRyy}$$

which is 1 in 16 (as in the Punnett Square).

Recombination

In Mendel's dihybrid cross of round yellow × wrinkled green we have already mentioned that two of the F_2 phenotypes resembled the original parents (round yellow and wrinkled green) and that two of them displayed new combinations of characters that were not shown by either of the parents (round green and wrinkled yellow). *The process by which new combinations of parental characters may arise* is known as **recombination**, and the individuals which possess the new combinations are called **recombinants**. Recombination is important because it is one of the factors which leads to genetic variation in natural populations—which we will have more to say about in Chapters 7 and 18.

The term 'recombinant' refers to the phenotype of an individual and recombinants are always classified with respect to the parental types originally used in the cross. In the cross round yellow × wrinkled green for example, the two pairs of characters are combined into the F_1, which resembles the maternal parent (because of dominance), and then *recombined* during gamete formation in the F_1 to give the two new phenotypes in the F_2 (Fig. 5.6).

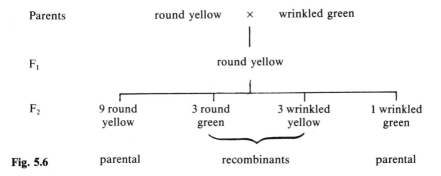

Fig. 5.6

Parents   round yellow   ×   wrinkled green

F_1 round yellow

F_2 9 round yellow   3 round green   3 wrinkled yellow   1 wrinkled green

parental   recombinants   parental

Importance of Mendel's work

The importance of Mendel's work received recognition from his successors who redefined it in their own terms and gave us what are now known as **Mendel's Laws of Heredity**. The 'Laws' are also referred to as the **First and Second Principles of Genetics**.

Why is Mendel's work so important?

The answer is straightforward. In the first place he gave us the method of genetic experimentation and explained how heredity works. Segregation simply means that a heterozygote, **Aa**, produces two kinds of gametes in equal frequencies of $\frac{1}{2}A + \frac{1}{2}a$. Pure-breeding homozygotes do not segregate; they only produce one kind of gamete: **AA** → all **A**, **aa** → all **a**. When two or more genes are involved in a cross they segregate independently. If an individual is heterozygous for both gene pairs, **AaBb**, we get four classes of gametes in equal frequencies—$\frac{1}{4}$**AB**, $\frac{1}{4}$**Ab**, $\frac{1}{4}$**aB**, $\frac{1}{4}$**ab**. When only one of the two pairs is heterozygous then only that pair segregates and we get two classes of gametes in equal frequencies—**AaBB** → $\frac{1}{2}$**AB**, $\frac{1}{2}$**aB**. Using these basic rules for one or more pairs of genes it is a simple matter to obtain the kinds and frequencies of gametes for whatever pair of individuals are involved in a cross, and then to find their random combinations by the Punnett Square method (the rules only apply to unlinked genes on different non-homologous pairs of chromosomes).

In the second place people could now understand the nature of heredity, whereas previously they had no proper idea about it. Before 1900 there was a widely held view that the gametes contained 'essences' which were derived from cells of various parts of the body, and that when fertilisation occurred they were blended together in some way to give the form of the progeny. **Blending inheritance** was held to explain how offspring could show some of the characters of both of their parents, but it was inconsistent with the observation that characters could turn up in the later generations of a cross in an 'undiluted' form. Mendel's work gave rise to a completely novel idea, and to what is probably the most important *concept* ever formulated in genetics—that of **particulate inheritance**. According to this idea, it is not the character itself which is transmitted during reproduction, but some discrete particles within the cells which determine each of the characters. These particles maintain their identity and separateness during inheritance through a hybrid and are handed on from one generation to the next in an unmodified and constant form. In anticipating the existence of such particles Mendel discovered the units of heredity which Johannsen later (1909) christened '**genes**'. His method of experimentation also laid the foundation of what Bateson was later to call the science of '**genetics**'.

Box 5.2   Gregor Johann Mendel (1822–1884)

Mendel was born into a poor peasant family in a small village in Moravia (now part of Czechoslovakia) in 1822. The poverty endured by himself and by his family placed Mendel under considerable strain throughout his early education, first in the village school and later in the Gymnasium in Troppau, and he suffered repeated illnesses. Despite several breakdowns, Mendel held a position near the top of most of his classes and in 1841 he went on to College in Olmuetz to study philosophy. Further breakdowns followed and by the time of his graduation in 1843 he finally abandoned his studies to take up holy orders and be admitted as a novice in the Augustinian convent at Brünn. He was given the name of Gregor, which he thereafter used in front of his baptismal name of Johann, and was assigned duties as the assistant teacher of physics in the school of Brünn. However, he failed in the examination for a regular teaching licence. He was then sent by his order to the University of Vienna for four terms to study natural sciences and mathematics, and after a brief spell back in the convent he returned there in 1853 to resit the examination, but failed again.

Mendel remained a teaching assistant for the next 14 years, and it was during this period that he carried out most of his research. Throughout this time he lived in two rooms which he turned into a laboratory, keeping a range of animals including birds, grey and white mice, and even at one stage a porcupine and a fox. In the garden he kept bees, took daily meteorological recordings and began his experiments with plant hybridisation.

In 1865 he presented his two papers at a meeting of the Brünn Society for Natural Sciences. According to Iltis, the audience 'listened with considerable astonishment . . . the attention of the hearers was inclined to wander when the lecturer was engaged in rather difficult mathematical deductions'. His account was published in the 1865 volume of the proceedings but little attention was paid to it. In a letter to the botanist Nägeli (April 18th, 1867) Mendel himself wrote about the result of his lectures: 'I encountered, and nothing else could (of course) be expected, very divided opinions; repetition of the experiments was, however, as far as I know, undertaken by nobody'.

The active life of scientific research and teaching which Mendel so much enjoyed was interrupted in 1868 by his election to the position of Abbot of the convent. He now directed his energies to management of the many properties which the convent possessed and to the pleasures of office. He entertained many prominent guests at the convent which had a reputation for the quality of its food. This indulgence led him to put on weight which he was constantly trying to reduce by various means, such as rolling on the floor of his bedroom or rising at four o'clock in the morning, or adopting a liquid diet, none of which proved very effective. He also became accustomed to smoking heavily and towards the end of his life he was partaking of up to twenty cigars a day. He travelled abroad and in Brünn he was engaged in many administrative activities for the government as well as directing an institute for deaf-mutes and the government mortgage bank. These duties left him little time for research and in the last ten years of his life he became embroiled in lengthy arguments over taxation on religious properties. Then Mendel was taken seriously ill, and in 1884 he died at the age of 62 from several causes including heart failure, kidney disease and hydropsy. His fear of being buried alive led him to insist on autopsy.

The impression that emerges of Mendel is of a man with a kind, tranquil, friendly and humorous nature. Eichling, who met the Abbot Mendel in 1878, wrote—'My first impression was a genuine and pleasant surprise—My customer's account had built in my mind a picture of an old, wrinkled, spooky, monk. Coming towards me was a fine looking, spectacled priest, smiling and extending a welcoming hand. His countenance expressed both determination and kindliness.' Mendel was also a clear-minded and logical thinker. He was an able teacher and much beloved by his pupils in the Brünn school. His adherence to facts, his thoroughness and the simplicity and directness of his dealings also made him invaluable in the position of office, and of course led to the excellence of his experimental approach.

According to Kalmus, the ingenuity, novelty and success of his work can be attributed to his ability to draw together his experiences from three different

Fig. 5.7 Gregor Mendel. (Photograph courtesy of The Mansell Collection.)

scientific disciplines in which he had received a rigorous training during the hardships of his young and formative years: philosophy, mathematics and biology. The philosophy of Aristotle, concerning the notion of 'opposites' and 'one quality being the alternative to another', may well have planted in his mind the idea of pairs of alternative characters. His knowledge of algebra enabled him to understand the F_2 ratios and to see how they were the result of randomly combining pairs of cell elements, as derived by expansion of the binomial expression $(A+a)(A+a)$. The educational training he received, with its emphasis on logic and mathematics, meant that he would have taken for granted the systematic approach that led him to count the numbers of seeds and of plants with contrasting characters, and to observe their *ratio*, whereas others who preceded him had simply noted the 'mixture'. Since childhood Mendel had been interested in gardening, and in flowers, and had himself produced a variety of *Fuchsia.* It was this interest that led him to start his experiments with the then fashionable aim in mind of explaining the possible role of hybrids in the creation of new species.

   Mendel, who had led a full and successful life, was not in any way bitter about the contemporary lack of interest in his work: as Niessl, a friend of his, wrote— 'Mendel did not expect any thing better, but I heard him in the garden, among his cultures of *Hieracia* and *Cirsiums,* express the prophetic words: "Meine Zeit wird schon Kommen" ("My time will come")'.

Introducing *Drosophila*

The fruit fly, *Drosophila melanogaster* (Fig. 5.8), is one of the most intensively studied of all 'genetic organisms'. It completes its life cycle in the short duration of only 10 days at 25 °C, it is easy to handle in the laboratory and has many excellent characters which can be used to identify its genes. It has four pairs of chromosomes $2n = 8$.

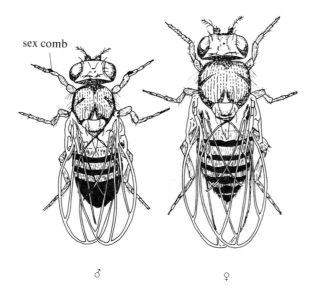

sex comb

♂   ♀

Fig. 5.8 Male and female specimens of the fruit fly—*Drosophila melanogaster*. The female is larger than the male and has lighter colouration on the abdomen. A diagnostic feature of the male is the sex comb (tuft of bristles) on the front leg.

Drosophila geneticists use internationally agreed names and symbols for the mutant forms of various characters. A **mutant** is *an inherited departure from the normal, or* **wild type**, *phenotype*. Each mutant character is given a name describing its characters; this is usually an adjective such as 'ebony' for ebony-coloured body, and a symbol which is an abbreviation of the name. The symbol may consist of the initial letter of the mutant name (e.g. **e** for ebony) or the initial letter plus an additional letter (e.g. **vg** for vestigial wing). Recessive mutants are represented by a small letter (e.g. **bw** for brown eye) and dominant mutants by capital letters (e.g. **B** for bar eye: eye reduced to a narrow slit). These same symbols are also used for the alleles that determine the characters; that is **e** stands for ebony body and for the allele that determines ebony body. The normal (wild type) allele at a particular locus is given the same symbol as the mutant allele, but with a '+' superscript. For the ebony body locus we therefore represent the genotypes and phenotypes as follows:

$$e^+e^+\text{—normal grey body}$$
$$e^+e \text{ —normal grey body}$$
$$e \ e \text{ —ebony}$$

All of the *Drosophila* mutants we shall be referring to in this book are recessive.

A description of some well known mutant characters in *Drosophila*, together with their names and symbols, is given in the table below:

Table 5.1   Examples of mutant characters in *Drosophila*

Symbol	Name	Description
w	white	white eye colour
m	miniature	small wings
bw	brown	brown eye colour
vg	vestigial	vestigial wings
e	ebony	ebony body colour
B	Bar	eye reduced to a narrow slit

Example of a dihybrid cross in *Drosophila*

A pure-breeding female fruit fly with the normal phenotype of long wings and grey body was mated with a pure-breeding male fly with vestigial wings and ebony body. The F_1 were all normal in appearance and when mated together gave an F_2 with four phenotypic classes in the proportions shown in Fig. 5.9.

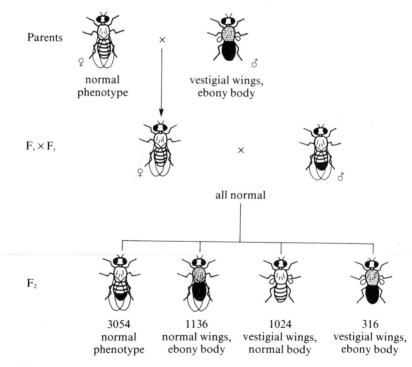

Fig. 5.9   The result of an experiment in which a wild-type *Drosophila*, with normal length wings and grey body, was crossed to a mutant male with vestigial wings and ebony body. What can we deduce about the inheritance of the two characters 'wing form' and 'body colour'?

How can we explain these results in terms of the inheritance of the characters 'wing form' and 'body colour'?

The F_1 has the wild type characters and resembles the female parent. We deduce that the characters vestigial and ebony are recessive. In the F_2 there are four phenotypic classes which are suggestive of a $9:3:3:1$ ratio—an indication that this is so can be derived by dividing each class by the lowest one, which is 316. Expected numbers for a $9:3:3:1$ from a sample of 5520 F_2 flies are $3105:1035:1035:345$, which fit the observed numbers very closely.

The interpretation that we make, from our knowledge of Mendel's second law, is that each of the characters is determined by a single gene with two alleles and that in the gametes of the heterozygous F_1 the two pairs of alleles are segregating independently of one another (Fig. 5.10).

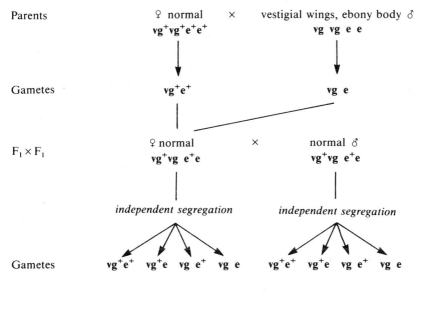

Fig. 5.10

In this example it is quite obvious that the observed numbers of the four classes of F_2 phenotypes agree closely with those which we would expect for a $9:3:3:1$ ratio. But this may not always be the case. We may be uncertain in some situations as to whether an observed set of experimental results do, or do not, fit a particular genetic ratio. To help us make a decision we use the statistical test outlined in Box 5.3.

Box 5.3   Chi-squared test

As we have explained earlier (Box 4.2), chance events are involved in the processes of heredity, and experimental results are subject to sampling error. This means that the numbers of individuals we actually observe in the various phenotypic classes of our F_2, or testcross progenies, will deviate to some extent from those which we expect to find. If the deviations between the observed and expected values are large there may be some difficulty in deciding whether the results obtained agree with the ratio concerned or not. Mendel clearly recognised this problem and overcame it by working with large samples of F_2s—more than 8000 in one case (Box 4.1). He had to do this because there were no statistical methods available at the time for assessing the reliability of his data.

The problem is straightforward. Consider the progeny from a cross between two red-flowered plants, $Rr \times Rr$, where R is the dominant allele which determines red flowers and rr gives white. We expect a $3:1$ ratio of red:white flowered plants. If we observe 152 red and 48 white, we will be quite satisfied with the result (the expected values being 150 and 50 for a sample of 200). But suppose we find, say (a) 156 red and 44 white, or (b) 165 red and 35 white. Is either, or both, of these outcomes acceptable as a $3:1$ ratio? The answer is that 156 red and 44 white is in agreement with our hypothesis of a $3:1$ ratio, but 165 and 35 is not. We arrive at this conclusion by the use of a simple statistical procedure known as the **chi-squared test** (χ^2 test). Essentially the χ^2 value which we calculate is a measure of the size of the difference (deviation) between the **observed** and **expected** results. The test takes account of the size of the sample, as well as the deviation, and gives the answer as a single numerical value. We use this value to assess how likely it is that the deviations we observe can be accounted for by *chance*, or whether they represent a *real departure* from the expected ratio.

The formula for calculating chi-squared is:

$$\chi^2 = \sum \frac{(O-E)^2}{E}$$

where $\sum$ = sum of; O = observed values; E = expected values.

In the example (a) given above the observed values are 156 and 44. The total size of the sample is 200 plants. The expected values for a $3:1$ ratio are therefore 150 and 50 (i.e. $\frac{3}{4} \times 200$ and $\frac{1}{4} \times 200$). The step-by-step procedure for working out the χ^2 value is as follows:

	O	E	Deviation (d) $O-E$	d^2	d^2/E
red	156	150	+6	36	0.24
white	44	50	−6	36	0.72
	200	200	0		$\chi^2 = 0.96$

We now consult a table of the distribution of χ^2 (Table 5.2) to find out what the probability (P) is of obtaining a deviation as large as, or larger than, the one we have *by chance alone*. The table takes account of the fact that the size of the χ^2 depends upon the number of independent comparisons (phenotypic classes) which contribute to its value as well as the size of the deviations in the individual classes (in this case two classes, red and white). If we are summing over four classes, as in the third example (below), then the χ^2 will be larger simply because we are adding together four numbers rather than two. We allow for the number of independent comparisons involved in our test by using the **degrees of freedom** (d.f.) which appear down the left-hand side of the table. Two classes (red and white) have only one degree of freedom because there is only one independent comparison, i.e. given the number in *one* class the number in the other class is fixed for a constant sample size. The number of d.f. for the tests described here is always one less than the number of classes so that, for example, four classes have 3 d.f. In the present example (a), therefore, we have one d.f. and we use the top line in the table. We notice that our χ^2 value of 0.96 is less than 1.074 for $P = 0.30$ (or 30%), but more than 0.455 for $P = 0.50$ (or 50%). This tells us that if there were no real deviations from a 3:1 ratio, and we repeated the experiment a large number of times, deviations as great as, or greater than, the one we have here (± 6) would occur in more than 30% of trials due to chance alone. In other words there is a high probability that the deviation which we obtained in our experiment is due to chance, and we can confidently conclude that there is no significant departure from a 3:1 ratio. Our results agree with expectations. The level of probability at which we decide to accept or reject our hypothesis of agreement with a given ratio is usually taken at 5% ($P = 0.05$). Any value of χ^2 smaller than 3.841, for one degree of freedom, is considered to be **non-significant** and the observed values agree with expectations.

Table 5.2   Distribution of χ^2

Degrees of freedom	Probability (P)								
	0.99	0.95	0.80	0.70	0.50	0.30	0.20	0.05	0.01
1	0.00016	0.004	0.064	0.148	0.455	1.074	1.642	3.841	6.635
2	0.0201	0.103	0.446	0.713	1.386	2.408	3.219	5.991	9.210
3	0.115	0.352	1.005	1.424	2.366	3.665	4.642	7.815	11.341

[From Fisher and Yates, *Statistical Tables for Biological, Agricultural and Medical Research*, Oliver and Boyd. Reproduced by permission of the authors and publishers.]

Now consider the other set of data given in example (b) above, where we have red and white-flowered classes of 165 and 35 respectively. The χ^2 value in this case works out at 6.0. For one degree of freedom it falls between the probability levels of $P = 0.01$ (or 1%) and $P = 0.05$ (or 5%). Because $P < 0.05$ we conclude that the χ^2 value is **significant** at the 5% level. The probability that a deviation of this size is due to chance alone is very low—less than 1 in 20—and we take this to mean that our hypothesis was incorrect, and that the data do not fit the expected 3:1 ratio.

As a third example we will use Mendel's results for the testcross **RrYy × rryy** described on p. 56. The hypothesis we wish to test is that the data agree with a $1:1:1:1$ ratio:

Phenotypic class	O	E	d	d^2/E
round yellow	55	52	+3	0.173
round green	51	52	−1	0.019
wrinkled yellow	49	52	−3	0.173
wrinkled green	53	52	+1	0.019
	208	208	0	$\chi^2 = 0.384$

There are now four classes in our test and we use the row in the table that corresponds to 3 d.f. Our χ^2 value falls between the probability levels of $P = 0.80$ (or 80%) and $P = 0.95$ (or 95%), and is therefore non-significant. There is a greater than 80% probability that the deviation from a $1:1:1:1$ ratio is due to chance, i.e. there is no real departure of the observed from the expected values and the data agree with our hypothesis.

The chi-squared test can be used to analyse any genetic ratios ($3:1$, $1:1$, $9:3:3:1$, $1:1:1:1$) including those which will appear in some of the forthcoming chapters (7, 8, 9 and 10). It is a valuable statistical device for assessing the reliability of data and for helping us to make decisions as to whether our data do, or do not, agree with a hypothesis that we have put forward. Where results do not conform with expectations it is then another matter to decide upon the possible reasons for the lack of agreement.

Summary

In Mendel's experiments, two or more pairs of contrasting characters were found to be inherited independently of one another. He concluded that this happened because each character was controlled by a different pair of cell elements, which we now call alleles of a gene. The two pairs of alleles segregate independently of one another so that the heterozygous F_1 dihybrid gives four kinds of gametes in equal frequencies. Random combination of the gametes, from male and female F_1s, gives a $9:3:3:1$ ratio of phenotypes in the F_2. The experiments on dihybrids gave rise to what is now known as Mendel's Second Law, or the Law of Independent Segregation.

Further reading

Kalmus, H. (1983), 'The scholastic origins of Mendel's concepts', *History of Science*, **21**, 61–83.

Lewis, K. R. and John, B. (1972), *The Matter of Mendelian Heredity*, Longman.

Mendel, G. (1866), 'Versuche über Pflanzenhybriden,' *Verhandlungen des Naturforschenden Vereins in Brünn*, **4**, 3–44.

Mendel, G. (1965), *Experiments in Plant Hybridisation* (Mendel's original paper in English translation with Commentary and Assessment by R. A. Fisher; together with a reprint of W. Bateson's Biographical Notice of Mendel), edited by J. H. Bennett, Oliver and Boyd.

Olby, R. C. (1966), *Origins of Mendelism*, Constable.

Parker, R. E. (1979), *Introductory Statistics for Biology*, Edward Arnold.

Questions and problems on Chapters 4 and 5

1 State Mendel's first and second laws (a) as originally stated, and (b) using modern terminology. How would you explain these two laws? Do you think we are justified in calling them 'laws'?

2 You have a single individual of dominant phenotype and you wish to determine whether it is homozygous or heterozygous for the allele concerned. Explain how you would do this if the organism is (a) a higher animal, (b) a cross-pollinating plant and (c) a self-pollinating plant.

3 In man there is a gene which, when homozygous, causes a condition called 'sickle-cell anaemia', so named because the red blood cells are abnormal and assume a sickle shape. Death usually occurs before adulthood. Heterozygous persons generally appear normal, but under low-oxygen concentrations their red blood cells may also become sickle shaped.

   A young woman, about to be married, had a brother who died of sickle-cell anaemia. Worried about the possible prospect of the condition occurring in her own children, she seeks medical advice. When samples of her blood are taken and placed in a low oxygen concentration, her red cells become sickled. However, her prospective husband's blood remains normal. What could you tell her about her children? What can you say about her parents? Explain your answers fully.

4 In human beings the ability to taste phenylthiourea depends upon the presence of a dominant gene.
   (a) Construct diagrams showing:
       (i) the genotype and phenotype of the parental generation, the F_1 generation, and the gametes of the cross between a homozygous 'taster' and a homozygous 'non-taster';
       (ii) the genotype and phenotype of the progeny of a cross between two genotypes of the F_1 type in (i).
   (b) Study Fig. 5.11. State, with reasons, the possible genotype of each person A, B, C, D and E.

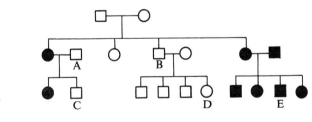

Fig. 5.11

○ taster female   ● non-taster female

□ taster male   ■ non-taster male.

5 A tall pea with red flowers was crossed with a tall pea with white flowers. The progeny had the following phenotypes:

   91 tall red; 97 tall white; 29 dwarf red; 33 dwarf white.

   What were the genotypes of the parents?

6 Tall, cut-leaved tomato plants are crossed with dwarf, potato-leaved plants giving in the F_1 generation nothing but tall, cut-leaved plants. These are allowed to cross with each other producing in the F_2 generation 926 tall, cut-leaved; 288 tall, potato-leaved; 293 dwarf, cut-leaved; and 104 dwarf, potato-leaved. What explanation can you offer for these results?

7 A red-flowered plant with the genotype **TtRr** is tall, but when selfed gives rise to seed which produces tall, red-flowered; tall, white-flowered; short, red-flowered and short, white-flowered plants in the ratio of 9:3:3:1, respectively. The short, white-flowered plants have the genotype **ttrr**.

The following cross was made:

$$\text{TtRr} \times \text{ttRr}$$

and may be represented by the type of diagram shown below:

	TR		tR	tr
tR	TtRR			ttRr
tr	TtRr		ttRr	

(a) Complete the empty boxes.
(b) Give the phenotypes for all genotypes from this cross in their correct ratio.

(O.L.E. Bot., 1980)

8 In the fruit fly *Drosophila* the wild type (normal) is grey in colour with wings that extend beyond the tip of the abdomen. Among the mutants of *Drosophila* are two which are respectively distinguished by dark body colour (ebony) and a vestigial condition of the wings (vestigial). A fruit fly with vestigial wings and ebony body colour is crossed to the wild type. The F_1 flies are backcrossed to the double recessive (ebony–vestigial) and the result is:

35 wild type; 32 ebony, normal wing; 33 vestigial, grey body colour; 34 ebony, vestigial.

Discuss this result with the aid of a diagram, commenting on the relationship between phenotypic appearance and genetic make-up. What would be the result of mating the wild types of this last experiment among themselves?

9 Black coat colour in cocker spaniels is controlled by a dominant allele **B** and red coat colour by its recessive allele **b**. Solid pattern is controlled by the dominant allele of an independently-segregating locus **S** and spotted pattern by its recessive allele **s**. A solid black male is mated to a solid red female who produces a litter of six puppies; two solid black, two solid red, one black and white and one red and white. Determine the genotypes of the parents.

10 Mendel's first law states that an organism's characteristics are determined by internal factors which occur in pairs and only one of a pair of such factors can be represented in a single gamete.
(a) Give the modern name for
   (i) an internal factor controlling part of an organism's characteristics,
   (ii) a pair of such factors.
(b) Explain why these factors occur in pairs in organisms rather than singly.
(c) Explain why only one of a pair of such factors can be represented in a single gamete.
(d) Rewrite Mendel's first law using modern genetic terminology.
(e) Explain what is meant by 'probability'. Give a simple example.
(f) In a repeat of one of Mendel's experiments a tall, coloured pea plant was crossed with a dwarf, white pea plant. The F_1 was self-fertilised and a sample of 20 seeds was collected from the F_2. None of these seeds carried the double-recessive characters, dwarf white. Explain why.

(*g*) What is the most probable number of dwarf coloured plants in a sample of twenty taken from the above experiment?

(*h*) Would a cross between parents, each carrying two factors, always produce four types of F_2 offspring? Yes or no? Explain your answer.

(W.J.E.C. Biol., 1979).

11 Kerry type cattle with normal leg length are produced by a homozygous genotype **DD** while short legged Dexter type cattle are known to possess the heterozygous genotype **Dd**. All those with homozygous genotype, **dd** are stillborn, very deformed and are known as 'bulldog calves'. The presence of horns in these cattle is governed by a recessive gene, **p**, the polled (hornless) condition is produced by its dominant allele **P**.

(*a*) State the kinds and proportions of gametes produced by polled (hornless) Dexter type cattle of the genotype **DdPp**.

(*b*) (i) Construct a diagram to determine the genotypes of the offspring produced by a cross between polled (hornless) Dexter type cattle of genotype **DdPp** and horned Kerry type cattle.

(ii) List the genotypes and corresponding phenotypes of the offspring.

(*c*) A farmer has both Kerry and Dexter cattle. In order to obtain Dexter calves he may either cross Kerry with Dexter or interbreed the Dexter cattle. He is advised that the Kerry with Dexter cross is the better breeding programme. Explain why this is so.   *(L. Biol., 1983)*

6
Genes and chromosomes

Now that we have studied the way in which the nucleus behaves during cell division, and the way in which character differences are inherited in breeding experiments, we are in a position to see the parallel between these two aspects of the process of heredity. It is obvious to us now, as it was to Sutton in 1903, that if the genes are actually part of the chromosomes then the Mendelian principles of segregation and independent segregation can be readily explained by the way in which the chromosomes behave during the reduction division of meiosis. In this chapter we will discuss the relationship between genes and chromosomes and develop the idea of the **Chromosome Theory of Inheritance**. The theory will be summarised at the end of Chapter 7 after we have also dealt with the principle of 'linkage'.

Relationship between genes and chromosomes

The simplified diagram in Fig. 6.1 shows the relationship between the genes and the chromosomes. In diploid organisms the chromosomes exist in homologous pairs. There are therefore two sets of chromosomes present in

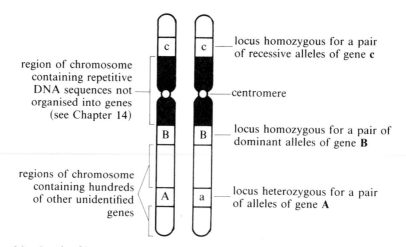

region of chromosome containing repetitive DNA sequences not organised into genes (see Chapter 14)

locus homozygous for a pair of recessive alleles of gene **c**

centromere

locus homozygous for a pair of dominant alleles of gene **B**

regions of chromosome containing hundreds of other unidentified genes

locus heterozygous for a pair of alleles of gene **A**

Fig. 6.1 A pair of homologous chromosomes shown in an unreplicated form, as single chromatids, as they would be at the end of somatic cell division. Three pairs of different genes are shown at different loci as well as some other features. A **locus** *is the site in the chromosome where a gene is located*: a particular gene is always found at the same locus. Homology means that a pair of chromosomes have alleles of the same genes, and other structural features, at corresponding places along their length.

the nucleus of each cell, one set originating from the mother (maternal set) and one from the father (paternal set). Each member of a homologous pair carries one of the alleles of each gene at corresponding positions called **loci** (single = **locus**) along its length. In a pure-breeding homozygote the homologues carry identical dominant (**BB**), or recessive (**cc**), alleles at the loci concerned. In a heterozygote (**Aa**) they carry a dissimilar pair—each homologue carrying one of the two alleles.

Meiosis and segregation

Segregation of a pair of alleles (**Aa**) takes place at meiosis because the genes are part of the chromosomes and are separated from one another when the chromosomes move apart at the anaphase stages. For a particular pair of alleles the segregation may take place when the homologous chromosomes move to opposite poles at anaphase I, in which case it is referred to as **first division segregation**, or when chromatids are separated at anaphase II, when it is called **second division segregation**. The difference between the two simply depends upon whether a crossover takes place between the gene locus concerned and the centromere, as explained in Fig. 6.2.

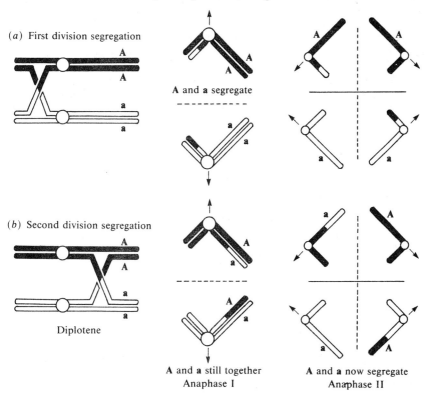

Fig. 6.2 Chromosomal basis of segregation for a pair of alleles at a single locus. The diplotene bivalent is shown with a single chiasma. When there is *no* crossing over between the gene and the centromere (*a*) the alleles segregate into different cells at AI. When crossing over does take place between the gene and the centromere (*b*) segregation is delayed until the second division (AII).

Normally, of course, we cannot tell at which stage of meiosis segregation takes place. We are unable to see the genes in the chromosomes and we have to deduce what is going on from our knowledge of meiosis and from the appearance of phenotypes in testcrosses or F_2 progeny. Ideally we would like to look directly at the haploid cells, or gametes, coming from individual meioses—if we could find a suitable character. As it happens, the ascomycete fungi provide just the right kind of material to enable us to observe segregation in individual cells. The haploid products of each meiosis are the ascospores, which are held together in clusters within the ascus. The details of how we use them are given in Box 6.1.

Box 6.1   Segregation in ascomycete fungi

The ascospores in these fungi are the haploid products of meiosis. We can see them quite easily down the microscope (Fig. 6.5) and by making use of mutations which affect their pigmentation we have a phenotype with which we can study the process of segregation. To see how this is done we must first remind ourselves of the details of the life cycle and sexual reproduction which were given for *Sordaria brevicollis* in Chapter 3. The details are virtually the same in two other well known species, *Sordaria fimicola* and *Neurospora crassa*.

Sordaria is a haploid organism and has only one set of chromosomes in each nucleus. When different mating types are present in a culture, hyphal fusion takes place and pairs of haploid nuclei come together in **binucleate cells** at the start of sexual reproduction. From these binucleate cells the fruiting bodies with their asci and ascospores are eventually formed.

From our point of view, what is of special interest is the way in which a single ascus develops from one diploid ascus initial cell as it undergoes meiosis (Fig. 6.3). When the asci are mature, each one contains the products of a single meiosis

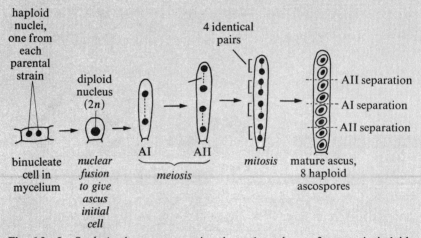

Fig. 6.3   In *Sordaria* the ascus contains the cell products of *one* meiosis laid out in order in a single row. The ascus develops in such a way that the arrangement of the ascospores can be related to the first and second divisions of meiosis. The nucleus which gave rise to the four spores in the top half of the ascus was separated from the nucleus giving the four spores in the bottom half of the ascus at anaphase I. Likewise the AII separation of nuclei gave the two pairs in each half of the ascus. The life cycle of *Sordaria* was given in more detail in Chapter 3.

and the spores are arranged in four pairs. They are present in the linear ascus in such a way that the four in the top half were separated from the four in the bottom half by the AI division, and the two pairs within each of the halves were separated at AII. In other words the ascospores are 'ordered' with respect to the first and second division of meiosis.

When a mutant strain having light-coloured buff spores (**b**) is crossed with a normal wild strain having black spores (**b$^+$**), then the diploid nucleus at the start of meiosis is heterozygous for the pair of alleles controlling ascospore colour. Because the ripe spores are ordered within the ascus, the colours act as 'markers' and enable us to see how the alternative forms of the gene were distributed at the two nuclear divisions (Figs 6.4 and 6.5). We cannot see the chromosomes in these fungi, because they are too small, but we can interpret the cytological events from our knowledge of chromosome behaviour in other species, and the

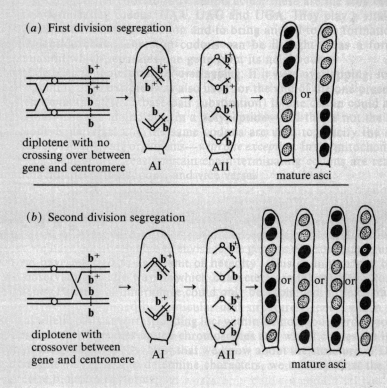

Fig. 6.4 Demonstration of segregation in *Sordaria brevicollis* using the buff-coloured ascospore mutant. To obtain meiosis in a heterozygote the wild-type black-spored strain (**b$^+$**) is mated with a buff mutant (**b**) by growing opposite mating types together on the same petri dish of agar. This is a direct demonstration of segregation because there are 4 black and 4 buff haploid spores, from the same meiosis, in each ascus. The stage of meiosis at which segregation occurs is shown by the spore pattern which is either (*a*) 4-4, or (*b*) 2-2-2-2. The two types of 4-4 asci are due to random orientation of the bivalent on the spindle at metaphase I: it is therefore a matter of chance whether the **b$^+$** alleles go to the top or bottom of the ascus. In the 2-2-2-2 asci the bivalent at MI, and the half-bivalents at MII, also attach to the spindle in a random orientation so that the four types are found in equal frequencies in a sample.

spore patterns can be readily explained. Segregation of a heterozygous pair of alleles from *one* meiosis is shown to give the exact Mendelian expectation of two nuclei (two pairs) carrying the one allele and two the other. The actual segregation process is deduced to take place at either AI or AII depending upon whether a crossover takes place between the gene and the centromere, or not (Fig. 6.4).

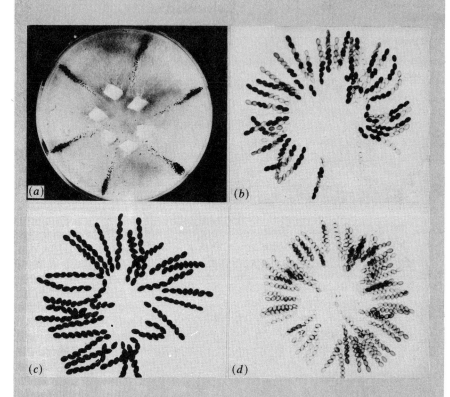

Fig. 6.5 Photos of fruiting bodies and ascospore formations in *Sordaria brevicollis.*

(*a*) Culture plate with fruiting bodies (perithecia). The plate was inoculated at the centre alternately with mycelium of plus and minus mating types. Sexual reproduction has taken place along the radii of the culture where mycelium of opposite mating types has grown together.

(*b*) Squashed out contents of a hybrid perithecium. The asci shown are from the cross of a normal black-spored strain × a mutant with buff-coloured ascospores. All of the different spore arrangements referred to in the text, and Fig. 6.4, can be seen. In a couple of asci the spores are out of order due to spindle slippage at the mitosis which follows meiosis.

(*c–d*) Asci from cultures of the two parents used to make the cross shown in (*b*). The black-spored strain is shown in (*c*) and the buff strain in (*d*). The variations in colour intensity in (*d*) are due to differences in the stage of maturity of asci.

Meiosis and independent segregation

The independent segregation of two or more pairs of characters, according to Mendel's second law, takes place because the genes controlling the characters are located in different non-homologous chromosomes. The chromosomal basis of this second principle of genetics is simply the random way in which chromosomes attach to the spindle at the metaphase stages. An explanation of the mechanism is given for two pairs of alleles, **AaBb**, in the simplified diagram in Fig. 6.6.

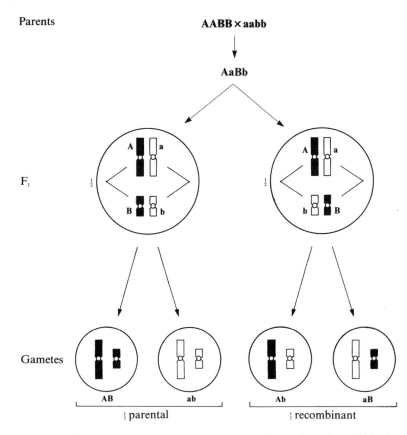

Fig. 6.6 Simplified diagram explaining how the random orientation of bivalents on the spindle at MI of meiosis accounts for the independent segregation, and recombination, of pairs of genes on different chromosomes. Chromosomes are represented as single structures without showing their chromatids.

The random orientation of bivalents with respect to one another leads to the production of four kinds of haploid gametes in equal frequencies. In dealing with independent segregation we do not need to concern ourselves with the consequences of crossing over within the individual bivalents. The effect of crossing over between the gene and the centromere is the same as described above for segregation, but the important factor here is the random orientation of the bivalents and it makes no difference to the overall outcome whether their individual pairs of alleles segregate at AI or AII.

Independent segregation and recombination

The independent segregation of genes in different chromosomes leads to recombination. Of the four kinds of gametes produced, when two pairs of alleles are segregating, two of them (**AB**, **ab**) are parental and two (**Ab**, **aB**) are recombinant (Fig. 6.6). This is known from the progeny phenotypes in the testcross, **AaBb × aabb**, and in the F_2 from the cross **AABB × aabb**.

When three pairs of genes are segregating independently, eight different combinations of the maternal and paternal chromosomes which carry them are found with equal frequency in the gametes. Two of the combinations are parental types and six are recombinant (Fig. 6.7).

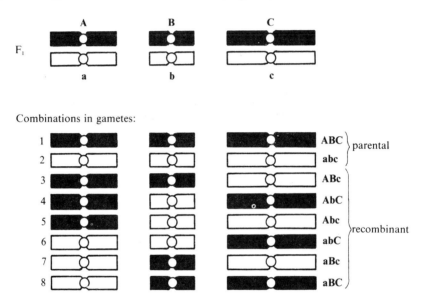

Fig. 6.7

In general terms, and leaving aside any complications due to crossing over, the number of different combinations of maternal and paternal chromosomes that can occur in a sample of gametes by independent segregation is 2^n, where $n =$ the number of chromosome pairs. For two chromosomes shown in Fig. 6.6 there are 4 possible kinds, $2^2 = 2 \times 2 = 4$. For three pairs there are eight, i.e. $2^3 = 2 \times 2 \times 2 = 8$, and so on. Each additional pair of chromosomes *doubles* the number of combinations. The chromosome number of man is $2n = 2x = 46$, i.e. 23 pairs; therefore it is possible for an individual male to recombine his maternal and paternal chromosome sets, by independent segregation, to give $2^{23} = 8\ 388\ 608$ genetically-unique sperms. Females have the same potential but they produce a relatively small number of egg cells during their lifetime.

We will discuss these matters of genetic variation further at the end of the next chapter as well as in some of the subsequent ones. For the time being we will take note of the fact that the basic chromosome number of a species has some genetical significance. The larger the number of different chromosomes into which the genetic material is divided the greater are the possibilities for recombination by independent segregation, and vice versa.

Summary

Genes are part of the chromosomes and the laws of heredity can be easily explained in terms of the way in which chromosomes behave when they undergo their reduction division at meiosis. Segregation is accounted for by the separation of homologous pairs at anaphase I, or by the movement apart of the daughter chromosomes at AII when there has been a crossover between the gene concerned and the centromere. Independent segregation is due to the fact that the two or more pairs of alleles are located in different non-homologous pairs of chromosomes which orientate at random on the spindle at metaphase.

Independent segregation gives rise to recombination and to genetic variation.

Questions and problems

1 The meiotic products in many ascomycetes form a linear tetrad and, in the majority, meiosis is followed by a mitotic division. The mitotic products also remain in the positions in which they were formed. Consequently, eight spores in single file are eventually produced in each ascus. Figure 6.8 shows the types of asci produced by a hybrid between a white spored and black spored strain of *Sordaria fimicola*. How would you account for the various spore arrangements?

Fig. 6.8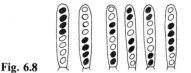

2 *Given:* a diploid organism heterozygous for three loci, as follows (chromosomes represented as a single line):

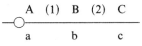

The centromere is shown just to the left of the A locus. By means of diagrams, showing the bivalent at diplotene and at the AI and AII stages, explain which genes segregate at AI of meiosis and which genes segregate at AII, with:
(*a*) A single chiasma in region (1) only,
(*b*) A single chiasma in region (2) only.

7
Linkage and recombination

The characters which Mendel studied showed independent inheritance because they were determined by genes in different chromosomes. If all chromosomes were made up of only one gene, then the laws of segregation and independent segregation would provide us with an adequate description of heredity. But this is not the case. We now know that the chromosomes are actually made up of linear sequences of large numbers of genes, which are all *linked* together and which cannot behave independently of one another in their inheritance.

In this chapter we will consider the inheritance of linked genes. We will also summarise the chromosome theory of heredity and the reasons why it is necessary to have meiosis at all.

Linkage

Linkage is *the association of certain genes in their inheritance.* The same term may also be used with reference to the characters determined by the linked genes. The concept of linkage can best be understood in the first place by describing an exceptional situation, and then proceeding to what is normal. In *Drosophila* the males have an unusual meiosis. There is no crossing over between homologous pairs of chromosomes, and therefore no chiasmata are formed in the prophase stages. At metaphase I the homologues simply lie side by side on the equator of the spindle and then separate from one another at anaphase I. Thereafter they behave normally. Because there is no exchange of chromatids between maternal and paternal partners in a bivalent, the genes in them show **complete linkage**.

Complete linkage

When pure-breeding male flies carrying the recessive mutations curled wings (symbol **cu**) and ebony body (e) are mated with normal females with straight wings (cu^+) and grey body (e^+) the F_1 are all normal and heterozygous for both pairs of alleles. If the F_1 *males* are now testcrossed with double recessive females, the four phenotypic classes which we might expect in the progeny, by independent segregation, are not present—instead there are only two: those with the parental combinations of cu^+e^+ and **cu e**. The dominant and recessive alleles segregate from one another in the F_1 as completely linked pairs, and not independently. The reason is that the two genes are in the same chromosome (i.e. the same pair of chromosomes) and not in two different ones as required for independent segregation. In the F_1 the two dominant alleles are together in the maternal chromosome and the two recessive alleles in the paternal partner of the bivalent (Fig. 7.1.).

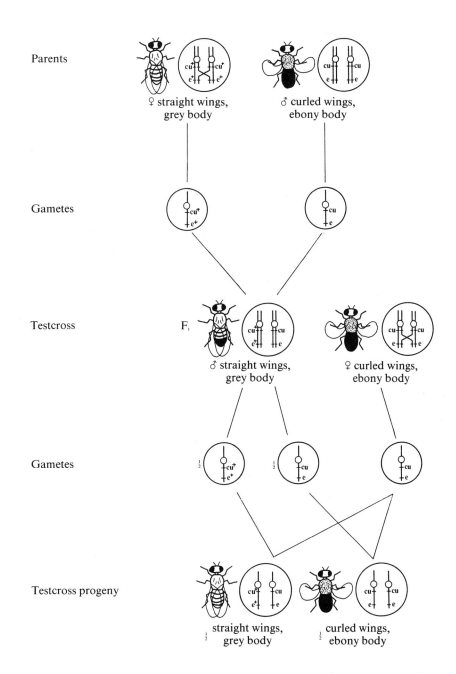

Fig. 7.1 Complete linkage in the fruit fly, *Drosophila melanogaster.* In the F₁ hybrid males there is no crossing over and the paternal and maternal homologues segregate unchanged into the gametes, taking the linked combinations of alleles with them. Consequently there are no recombinants and only two kinds of progeny are produced in equal numbers.

In dealing with linkage it is therefore helpful to change the notation slightly and to write the alleles one above the other, rather than side by side, and draw a line between them to represent the chromosomes (Fig. 7.2).

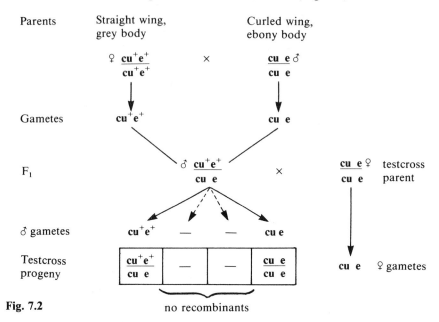

Fig. 7.2   no recombinants

Since there is no crossing over in the male, and no exchange of chromatid segments between homologous partners, the genes are completely linked and simply follow the chromosomes in their inheritance (Fig. 7.1). A bivalent without any crossing over behaves as in Fig. 7.3.

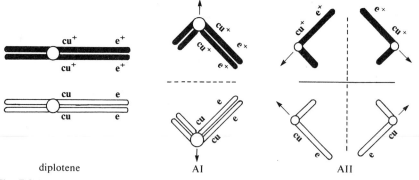

Fig. 7.3

Meiosis without crossing over and chiasma formation is not confined to male *Drosophila*. It is also found in some other insects, e.g. *Callimantis*, and in the plant species *Fritillaria japonica*. It is nevertheless an unusual situation and an extreme form of linked inheritance. Genes which are located in the same chromosome generally show **partial linkage**, because in normal meiosis there is always at least one chiasma formed somewhere within a bivalent, and this leads to a certain amount of recombination between the genes.

Partial linkage

If we consider the same cross again but this time using a *female* as the heterozygous F_1, and testcrossing her to a double recessive male, we get a different result (Fig. 7.4).

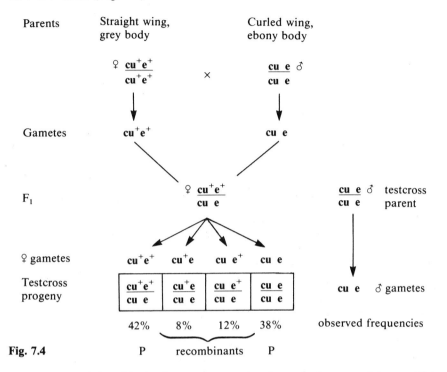

Fig. 7.4

We now have four kinds of progeny resulting from the four possible combinations of alleles in the F_1 gametes. Two of them are 'parentals' (P), resembling the two parent flies going into the cross, and two are recombinants possessing characters of both parents. We notice that they are not in the equal frequencies expected by independent segregation, and neither are they completely linked as they were when a male was used as the F_1 hybrid. This time we have partial linkage with the parental combinations making up the majority of the gametes produced by the F_1, and the minority being comprised of recombinants.

Crossing over

The reasons for this partial linkage have to do with the way in which the genes are arranged in the chromosomes and with the crossing over which takes place between them.

Crossing over is *a process of exchange between homologous chromosomes which gives rise to new combinations of characters*. The explanation is given in the diagram in Fig. 7.5. Crossing over gives rise to two kinds of recombinants in about equal numbers, because of the way in which two non-sister chromatids break and rejoin at corresponding sites within the bivalent during chiasma formation (Chapter 2).

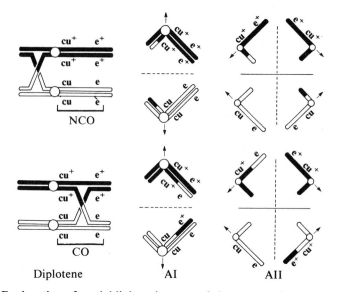

Diplotene   AI   AII

Fig. 7.5   Explanation of partial linkage in terms of chromosome behaviour at meiosis. Partially-linked genes are located in the same chromosome, and the parental combinations of alleles are transmitted together except in those cells where crossing over takes place between them. In a large sample of cells undergoing meiosis, crossing over will occur between the two genes in only some of the cells, depending upon how closely linked they are. In cells in which crossing over does take place, only two of the four chromatids are involved. For these reasons recombinants in a testcross comprise less than 50% of the progeny. (NCO = no crossing over between the two genes; CO = crossing over.)

When crossing over happens between the loci of the genes we are studying, two of the four haploid products of meiosis are recombinant and two are parental. When crossing over doesn't happen between the gene loci, the products of meiosis are all parental. Since a crossover can occur anywhere within a bivalent it will only take place between the two genes in some of the meiotic cells and not in others, and for this reason the recombinants usually make up much less than half of the testcross progeny. For any particular pair of genes the proportion of recombinants will always be about the same when the testcross is repeated (around 20% for **cu–e**), because the genes occupy fixed positions in the chromosomes and there is a certain probability of a crossover taking place between them. For a different pair of genes the proportion of recombinants will be different—as we will see later.

Coupling and repulsion

In the example used above the two dominant alleles (cu^+e^+) were contributed by the gamete from the female parent and the two recessives (**cu e**) by the gamete from the male. In the F_1 therefore the two dominants were on one chromosome and the two recessives on the other:

$$\frac{cu^+\ e^+}{cu\ \ e} \qquad coupling$$

This arrangement, in which the two dominant alleles are on one chromosome, and the two recessives are on the homologous partner, is known as **coupling**.

It is also possible to have an alternative situation where a dominant is associated with a recessive, i.e. where the alleles are linked in **repulsion**:

$$\frac{cu^+ \; e}{cu \; \; e^+} \quad \textit{repulsion}$$

For this arrangement to come about each parent has to contribute one dominant and one recessive allele, so the cross has to be made differently in the first place. The outcome is also different (Fig. 7.6).

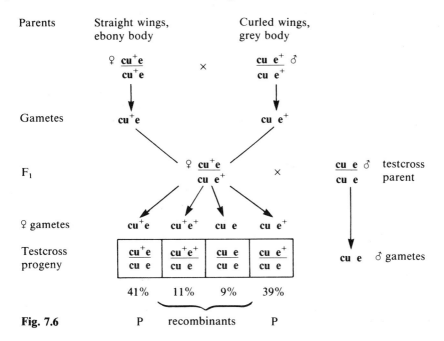

Fig. 7.6

As far as the phenotypes of the F_1 individuals are concerned it makes no difference whether the genes are linked in coupling or repulsion. In both cases they are heterozygous for the two loci and wild type. What distinguishes the two linkage arrangements is their breeding behaviour.

With coupling the recombinants will each carry one dominant and one recessive character, and we can identify them among the four classes of testcross progeny because they are the two classes with the smaller numbers, for reasons already explained. We can represent them, for convenience, in terms of the F_1 gametes (Fig. 7.7).

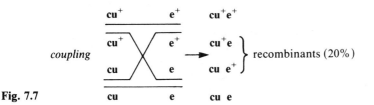

Fig. 7.7

In the case of repulsion it is the other way around, because the parents are different, and the recombinants now have associations of two dominants and two recessives (Fig. 7.8).

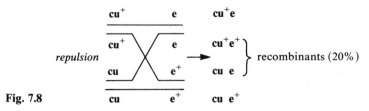

Fig. 7.8

This serves to remind us that we recognise a recombinant by reference to the combinations present in the parental form, rather than as a particular combination of alleles in the testcross progenies themselves.

Crossing over and recombination

It is important to realise that crossing over is only one of the two ways in which recombination can occur. The other way is by the independent segregation of unlinked genes (pp. 75-6).

Detection of linkage

How can we tell whether two genes are linked or not? The answer, as will be evident by now, is to make a testcross and to see whether or not the four classes of progeny are present in the equal frequencies expected by independent segregation. If they are *not* then the genes are linked. It is necessary to test for agreement with this expectation by making the appropriate statistical test (Box 5.3, p. 64).

There is a complication when we have two genes situated at opposite ends of a chromosome (Fig. 7.9).

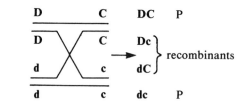

Fig. 7.9

In this case there will be at least one crossover between the two genes in every one of the cells undergoing meiosis, and half of the testcross progeny will be recombinant. Under these circumstances we cannot tell whether the genes are linked or not. The only way to demonstrate the linkage would be to use the male as the F_1 hybrid if we were dealing with *Drosophila*, or to have a third gene which would show linkage with both of the other two.

Box 7.1   Recombination frequencies and linkage maps

A crossover can take place anywhere within a bivalent. If two genes are at opposite ends of a chromosome it will happen between them in every cell, and half of the testcross progeny will be recombinant. If two genes are very close together, on the other hand, then crossing over between them will happen very rarely and there will be correspondingly fewer recombinants. *The percentage of a sample of testcross progeny that are recombinants* is the **recombination frequency** (or **crossover value**). To see how we calculate its value consider the following cross in *Drosophila* involving scarlet eyes (**st**) and curled wings (**cu**) (Fig. 7.10).

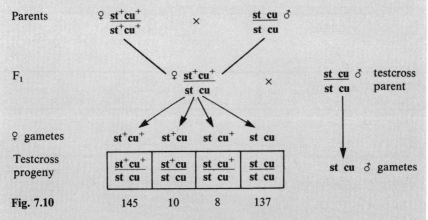

Fig. 7.10   145   10   8   137

We can represent the testcross progeny more simply in terms of their F_1 gametes because the double recessive **st cu** does not affect the phenotype of the progeny:

$$st^+cu^+ \qquad 145$$
$$st^+cu \qquad 10$$
$$st\ cu^+ \qquad 8$$
$$st\ cu \qquad \underline{137}$$
$$\Sigma\ \overline{300}$$

The parental combinations are **st⁺cu⁺** and **st cu**, and the recombinants arising by crossing over are the two smaller classes, **st⁺cu** and **st cu⁺**. Recombination frequency (or percentage) is calculated as follows:

$$\frac{\text{no. recombinants}}{\text{total progeny}} \times 100 = \frac{10+8}{300} \times 100 = 6\%$$

The recombination frequency gives us some idea of how far apart the two genes are on the chromosome. Genes at either end give a maximum of 50% recombination, so 6% must mean they are relatively close together. We have no *real* measurements of course, and our distances are therefore expressed in arbitrary map units. Geneticists have a standard system by which 1% recombination is taken as the equivalent of 1 map unit on a chromosome. The genes **st** and **cu** are therefore 6 map units apart—somewhere in the chromosome (Fig. 7.11).

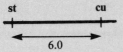

Fig. 7.11

Suppose we now make another testcross, involving the genes **st** and **e**, starting with the parents

$$\frac{st^+e^+}{st^+e^+} \times \frac{st\ e}{st\ e}$$

and we obtain the following set of data from the testcross:

$$st^+e^+ \qquad 112$$
$$st^+e \qquad 35$$
$$st\ e^+ \qquad 37$$
$$st\ e \qquad 116$$
$$\overline{\sum 300}$$

$$\text{Recombination frequency} = \frac{35+37}{300} \times 100 = 24\%$$

The genes **st** and **e** give a higher recombination frequency than **st** and **cu** because they are further apart in the chromosome and there is a greater probability that a crossover will occur between them. Their map distance is 24 units (Fig. 7.12).

Fig. 7.12

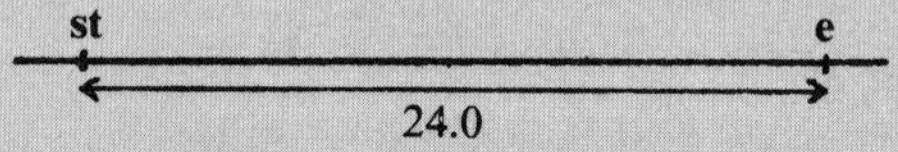

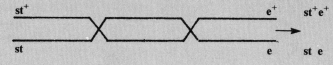

Where do we place **cu**? We know it is six map units from **st**. But is it between **st** and **e**, or to the left of **st**? From our earlier testcross results (p. 83) we know that **cu** and **e** give 20% recombination—**cu** is therefore between **st** and **e** (Fig. 7.13)

Fig. 7.13

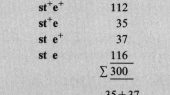

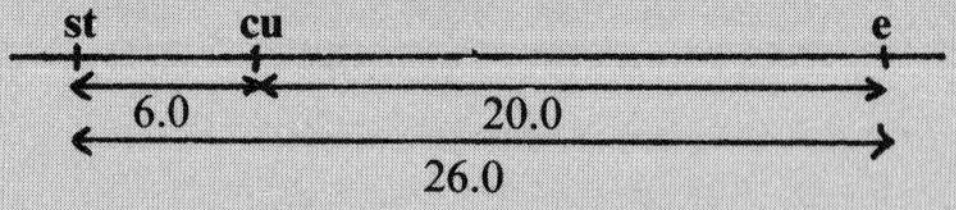

There is a discrepancy between the 24% recombination obtained from the testcross involving **st-e**, and the 26% by summing the distances **st-cu** and **cu-e**. This is because of a complication due to **double crossovers** which occur when genes are quite far apart and which are not detected as recombinants because they cancel one another out (Fig. 7.14).

Fig. 7.14

If we were working with the shorter distances these would be detected as single crossovers in the individual regions (Fig. 7.15).

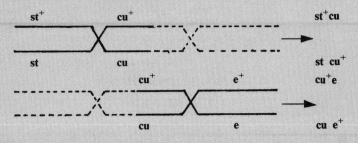

Fig. 7.15

The best estimate of the distance between **st** and **e** is therefore 26 map units—the sum of the two shorter distances.

These examples taken from mutants of chromosome III of *Drosophila* have shown us how recombination frequencies can be used to work out the order of genes along a chromosome and to make up a genetic linkage map. In some organisms where lots of genes are known (*Drosophila*, maize, tomato) very extensive chromosome maps have been compiled showing the order of genes within the different chromosomes and the relative distances between them. It turns out that there are as many linkage groups as there are chromosomes in the haploid set, and that longer chromosomes have more genes in them than the shorter ones. A simplified version of the linkage maps of *Drosophila* is shown in Fig. 7.16.

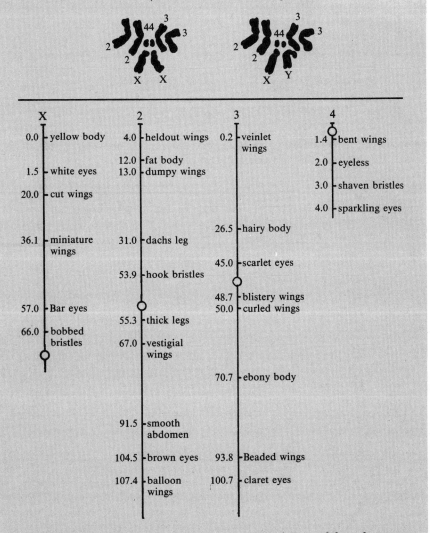

Fig. 7.16 Linkage map of the four chromosomes of *Drosophila melanogaster* showing a small selection of the numerous known gene loci.

Example of linked inheritance

Question:

In maize, coloured aleurone (**R**) is dominant to colourless (**r**). Yellow plant colour (**y**) is recessive to green (**Y**). Two plants, both heterozygous, were crossed to the double recessive, $\frac{ry}{ry}$, and gave the following progenies:

Phenotypes	Progeny of plant 1	Progeny of plant 2
Coloured aleurone; green plant	88	23
Coloured aleurone; yellow plant	12	170
Colourless aleurone; green plant	8	190
Colourless aleurone; yellow plant	92	17
	200	400

From the data: Calculate percentage recombination between **Y** and **R** and discuss the linkage between the genes in plants 1 and 2.

Answer:

In a testcross, the phenotypes of the progeny tell us what the genotypes were of the F_1 gametes from which they came, and it is useful to represent the progeny in terms of the gametes:

Plant 1			Plant 2		
RY	88		**RY**	23	(recomb.)
Ry	12	(recomb.)	**Ry**	170	
rY	8	(recomb.)	**rY**	190	
ry	92		**ry**	17	(recomb.)
	200			400	

In both cases the recombinants are the smaller classes (we are told the genes are linked), identified as (recomb.). The recombination percentage between **Y** and **R** (dominant symbols used only to name the genes) is as follows:

$$\frac{12+8}{200} \times 100 = 10\% \text{ for plant 1}$$

$$\frac{23+17}{400} \times 100 = 10\% \text{ for plant 2}$$

'Discuss the linkage between the genes'—this can only mean with reference to coupling and repulsion, because the two F_1s obviously differ in this respect. From the parental types (the two larger classes in each plant) we see that the genes are linked in coupling in plant 1, and in repulsion in plant 2:

plant 1 $\dfrac{R \ Y}{r \ y}$ plant 2 $\dfrac{R \ y}{r \ Y}$

The difference in linkage arrangement makes no difference to the recombination frequency between the two genes—because they are in the same position in the chromosomes in both cases.

Recombination and genetic variation

Significance of meiosis

In describing the principles of genetics, i.e. **segregation, independent segregation** and **linkage**, we have covered the basic facts about inheritance and given the rules, or laws, by which the transmission of genetic information takes place. We have also seen how these principles can be explained and understood once we realise that the genes are parts of the chromosomes and are carried with them during their complex processes of pairing, recombination and separation at meiosis. But why is it all necessary?

What is the significance of meiosis?

The answer is that during meiosis the genetic material in the maternal and paternal sets of chromosomes, which was brought together at fertilisation, is **recombined** and transmitted into the gametes in such a way that each sperm and egg produced has a unique set of genes (i.e. combinations of alleles). The progeny that are produced from crossing together two individuals of an outbreeding species therefore display genetic and phenotypic variability which results from the reshuffling of the genes in the sex cells of their parents. This reshuffling of genes is the function of meiosis and the advantage of sexual reproduction. The variation it generates is of fundamental importance to the long term survival and evolution of the species. Without genetic variability it would not be possible for natural selection to be effective and for a species to respond and to adapt to changes which may arise in its environment (see Chapter 20).

The processes of recombination, as we have seen, are two-fold: (1) independent segregation of genes in different chromosomes; (2) crossing over between genes which are linked.

The F_1 from parents which differ in two genes produces four kinds of gametes. With independent segregation they will be in equal frequencies. With linkage they will not: there will be an excess of the two parental classes and a deficiency ($<50\%$) of the recombinant ones. If the genes are very closely linked, the recombinant gametes may be quite rare, but they will still be found if a large enough sample is taken.

In both cases, because four kinds of gametes are produced, the F_2 will comprise nine genotypes and four phenotypes:

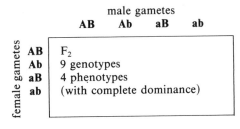

With independent segregation the phenotypes will be in the ratio $9:3:3:1$. With linkage (i.e. partial linkage) they will not be in any particular ratio, but they will all be present nonetheless.

If the parents differed in three genes, then recombination in the F_1 would lead to an F_2 with 27 genotypes and eight phenotypes. The table below shows examples of what happens as the number of allelic pairs by which the parents differ is increased:

No. allelic pairs	F_2 genotypes	F_2 phenotypes*
2	9	4
3	27	8
4	81	16
10	59 049	1024
21	10 460 353 203	2 091 152
n	3^n	2^n

*Where one allele is fully dominant over the other—this is not always the case, as explained in Chapter 9.

The effectiveness of recombination in bringing about genetic variation is self-evident.

In man it is estimated that there are of the order of 100 000 gene loci, distributed among the 23 chromosomes, and that on average about 7% of them are heterozygous. In other words an individual is heterozygous for some 6700 pairs of alleles and is therefore capable of producing 2^{6700} different kinds of germ cells. This is more than the number of atoms in the universe! Looked at in this way it is not surprising that every individual person in the world (except for monozygotic twins) is genetically unique.

Significance of diploidy

It follows from what has been said above that for meiosis to be effective in giving rise to genetic variation an organism must be sexually reproducing and have a diploid phase at some stage in its life cycle. Recombination can only occur when two sets of chromosomes which differ in the alleles of the genes they carry are brought together in the zygote during fertilisation. At meiosis, later on, the homologous sets will then pair and reshuffle their genes into the haploid spores or gametes which result from the reduction division process. Evolutionary advanced organisms have, presumably, found it beneficial to be diploid for the greater part of their life cycle and to have a very short haploid phase. This results in their carrying many alleles hidden in the recessive form which are of no obvious value, or may even be harmful, in the short term, but which provide a reservoir of genetic variation that is available for their long-term evolutionary needs. This aspect is discussed more fully in Chapters 18, 19 and 20.

Significance of linkage

Genes in different chromosomes assort quite freely and segregate their alleles into the gametes in all possible random combinations. Linked genes assort much less freely. The linking together of certain combinations of genes into the same chromosome gives some control over recombination, and means that certain combinations of characters can be kept together.

Chromosome theory of inheritance

The chromosome theory of heredity states that *the chromosomes are the carriers of the genes and represent the material basis of inheritance.*

The idea was first put forward by Sutton in 1903 who saw that the separation of homologous chromosomes at anaphase could provide the material basis for the separation of character differences during gamete formation as postulated by Mendel. Subsequently it was realised that chromosome behaviour during cell division could explain all the main aspects of 'nuclear inheritance' and the theory became firmly established. Now that we have dealt with the main aspects of inheritance it will be a useful stage at which to summarise the relationship between genes and chromosomes upon which the theory is founded:

1. Chromosomes contain the DNA which is the genetic material. There is one double helix in each chromatid (Chapter 13). Duplication and separation of chromatids during the mitotic cycle accounts for the regular way in which identical sets of genes are distributed to the daughter cells in somatic cell division.
2. Segregation is explained by the separation of homologous pairs of chromosomes at anaphase of meiosis.
3. Independent segregation comes about due to the random orientation of bivalents on the spindle at metaphase.
4. Linked inheritance takes place for genes which are located in the same chromosome. Gene mapping has shown that there are as many groups of linked genes as there are numbers of chromosomes in the haploid complement of a species. *Drosophila* has four linkage groups for its four chromosomes. Maize has ten linkage groups corresponding to its basic number of $x = 10$.
5. Recombination of linked genes can be understood in terms of crossing over between chromatids of homologous partners.
6. Modified patterns of sex-linked inheritance are explicable in terms of the sex chromosome differences between males and females. In *Drosophila* and in mammals each gene in the differential segment of the X chromosome is represented by one allele in males and two in females (Chapter 8).

Summary

Genes which are located in the same chromosome show linked inheritance. In an F_1 hybrid the combinations of alleles present in the maternal and paternal chromosomes of a bivalent are transmitted together into the gametes, except in those cells in which they are recombined by crossing over. The incidence of crossing over between linked genes depends upon their placement in the chromosomes. Those which are far apart will have more crossing over than those which are close together because the occurrence of a crossover in any particular site is a matter of chance.

By calculating the percentage of recombinants among a sample of testcross progeny it is possible to estimate the frequency of crossing over and to construct linkage maps showing the order of genes within a chromosome and the relative map distances between them.

The correspondence between the number of linkage groups and the number of different chromosomes, for a particular species, provides additional evidence for the chromosome theory of heredity. Crossing over between linked genes is one of the ways in which recombination takes place at meiosis. The other way is by independent segregation. The variation that comes about by recombining heterozygous combinations of alleles is the main function of meiosis and of the sexual system of reproduction.

Questions and problems

1 What is 'gene linkage'? Explain why linkage groups can be equated with chromosomes and show how recombination of genes on homologous chromosomes may occur. *(C. Bot., 1982)*

2 Explain what is meant by 'linkage' and 'crossing over'. Describe and explain breeding experiments you would perform, using a *named* organism, to find out whether or not two characters are linked. *(L. Biol., 1979)*

3 (*a*) What is the evidence upon which the chromosome theory of inheritance is based?
   (*b*) How does this theory help to explain (i) Mendel's 1st and 2nd Laws, (ii) linkage?
   (*c*) What is meant by 'complete linkage'? Explain what is different about the meiotic sequence in complete linkage.

4 Explain why no two gametes in the semen of a man are of the same genotype.

5 What are meant by the terms 'coupling' and 'repulsion'? Explain your answer fully with the use of clearly labelled diagrams. Discuss for both coupling and repulsion what is meant by the term 'recombinant'.

6 Explain the term 'crossover frequency'.
   In a certain organism genes **A**, **B**, **C** and **D** were studied and it was found that the crossover frequency between **A** and **B** was 20%, **A** and **C** 5%, **B** and **D** 5%, and **C** and **D** 30 per cent. What is the probable sequence of these genes on the chromosome and what would you expect the crossover frequency between genes **A** and **D** to be? Discuss the significance of crossing over.

7 Outline the main events during meiotic divisions in organisms. Discuss the ways in which meiosis generates variation in plants and animals. *(W.J.E.C. Biol., 1980)*

8 In the fruit fly *Drosophila* normal antennae and grey body are linked and are dominant to twisted antennae and black body.
   When a normal fly was crossed with one carrying the recessive alleles, the offspring were all of the normal type. One of these latter was then crossed with a fly homozygous for the recessive alleles and the following numbers of offspring were obtained.

Normal antennae and grey body	90	Twisted antennae and grey body	11
Normal antennae and black body	9	Twisted antennae and black body	86

Explain this result and make an annotated diagram of the bivalent which produced the recombinant classes.

9 A female fly, *Drosophila melanogaster*, with a normal (wild type) phenotype was crossed with a male fly of the following phenotype:

> dumpy wings (dp) as opposed to normal long wings (dp^+),
> short legs (s) as opposed to normal long legs (s^+),
> purple eyes (pr) as opposed to normal red eyes (pr^+).

The offspring of this cross were:

$dp\ s\ pr^+$	60
$dp^+\ s^+\ pr$	60
$dp\ s\ pr$	18
$dp^+\ s^+\ pr^+$	18
$dp\ s^+\ pr$	18
$dp^+\ s\ pr^+$	18
$dp\ s^+\ pr^+$	2
$dp^+\ s\ pr$	2

Comment on these results.   *(L. Zool., Special Paper, 1983)*

10 Bateson and Punnett crossed sweet pea (*Lathyrus odorata*) plants with purple flowers and long pollen grains with plants with red flowers and round pollen grains. The F_1 plants were all purple-flowered with long pollen grains. However, in the F_2 there were

4831	purple, long
390	purple, round
393	red, long
1338	red, round plants

a ratio of about $11:1:1:3$.
(*a*) What theoretical ratio would you expect the four classes to show in the F_2?
(*b*) How would you explain the experimental ratio?
(*c*) By what process could such new combinations as purple flowers and round pollen grains arise?   *(C. Biol., 1983)*

11 In the fruitfly *Drosophila melanogaster* vestigial wings (**vg**) and black body (**b**) are recessive autosomal genes. When vestigial-winged females were crossed with black-bodied males all the F_1 offspring were wild types. When males from this F_1 generation were crossed with vestigial-winged black-bodied females the offspring were:
  (1) 50% vestigial-winged, normal body colour;
  (2) 50% black-bodied, normal-winged.
(*a*) By means of a simple diagram show how this result was obtained.
(*b*) What genetic principle do these results illustrate?
(*c*) When females from the F_1 were crossed with vestigial-winged, black-bodied males the following result was obtained:
   (i) 40% vestigial-winged, normal body colour;
   (ii) 40% black-bodied, normal-winged;
   (iii) 10% wild-type;
   (iv) 10% vestigial-winged, black-bodied.
  What were the genotypes of these offspring?
(*d*) Explain the appearance of wild type and double-recessive flies in this cross.
                    (W.J.E.C. Biol., 1978)

8
Sex-linked inheritance

Sexual differentiation of a species into separate male and female forms is usually genetically determined. Geneticists soon realised that the inheritance of sex could be interpreted in Mendelian terms, since a mating between male and female always resulted in two sexes among the offspring in approximately equal numbers—as would be achieved in a monohybrid testcross. In some species sexuality is simply associated with a single gene difference; e.g. in the mosquito where the male is **Mm** and the female **mm**. More commonly, however, an examination of males and females reveals a difference in their chromosome complements, and sex determination is associated with the inheritance of particular chromosomes.

The term **sex chromosomes** is used to describe *the chromosomes in the complement which carry the sex-determining genes.*

All the other chromosomes in the set which are not involved in sex determination are known as the **autosomes**.

As the sex chromosomes are usually different in structure and genetic organisation in the two sexes, there are modified patterns of inheritance of the genes which are located in them. Indeed the association of modified patterns of inheritance with a particular pair of chromosomes within the complement provided some of the earliest and most convincing evidence for the chromosome theory of heredity.

In this chapter we will consider some of the well known sex chromosome systems in animals, and the modified forms of Mendelian inheritance that are linked to them. We will also discuss mechanisms of sex determination and some of the complications, in terms of gene action and expression, that arise from having genes carried in the sex chromosomes.

Sex chromosomes

The XO system

Early cytologists studying chromosomes in insects belonging to the order Orthoptera (grasshoppers and locusts) found that one member of the chromosome complement, which was paired in the female, had only a single representative in the male. This became known as the X chromosome. In the grasshopper *Chorthippus parallelus*, for example, the male has 17 chromosomes (including one X) and the female 18 (including two Xs). During meiosis in the female the two Xs pair together and behave in the same way as in any other bivalent. In the male the single X has no pairing partner: it moves undivided to one pole of the cell at anaphase I and then divides normally at AII. Half of the sperms therefore contain a single X and the other half do not (O).

Since the male produces two types of gametes, with respect to the sex chromosomes, it is referred to as the **heterogametic sex** (XO).

The female (XX) *produces only one kind of gamete* (containing an X) *and is therefore called the* **homogametic sex.**

The transmission of the sex chromosomes follows a simple pattern of inheritance, and the sex ratio of $\frac{1}{2}$ male : $\frac{1}{2}$ female is maintained in the progeny of crosses (Fig. 8.1).

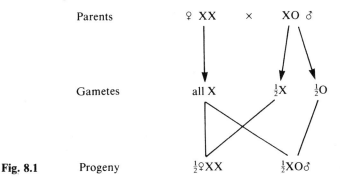

Fig. 8.1

The XX/XY system

In many species of flies, including *Drosophila*, and in many mammals, including man, there are two kinds of sex chromosomes. They are designated the X and the Y. Their morphology and relative size varies according to the species. The X may be the largest or, less commonly, the smallest member of the complement. It is usually larger than the Y, as in the case of man (Fig. 8.2), but is occasionally smaller as in *Drosophila* (Fig. 8.3). In flies and in mammals the female is the homogametic sex (XX), producing eggs carrying a single X, and the male is heterogametic (XY). At meiosis in the male the X and Y form a bivalent in prophase I and then segregate to opposite poles of the cell at AI. At anaphase II the chromatids separate so that half of the sperms contain a single X and the other half a Y. The sex of the offspring is therefore determined by the male parent, as illustrated below for *Drosophila*. In a large sample of offspring half will be male and half female. For an individual the probability of being male or female is $\frac{1}{2}$ in both cases (Fig. 8.4).

In butterflies, moths and birds the same sex chromosome system operates but the roles are reversed. The homogametic sex is the male and the heterogametic one the female.

Many different sex-determining systems operate amongst higher organisms. An interesting example is that found in the honey bee. In this species sex is determined by the ploidy level (Chapter 16) which is controlled by the female queen bee. The queen is diploid with $2n = 2x = 32$. By controlling the sphincter of her sperm receptacle she can lay either unfertilised eggs with the haploid (reduced) number of 16 chromosomes, or fertilised ones with 32 chromosomes. The fertilised diploid eggs develop into females, and the way in which they are fed determines whether they will be a worker or a queen. Unfertilised haploid eggs become males (i.e. drones). Since the male is haploid, with only one set of chromosomes, sperm production depends on a modified meiosis which is essentially a mitosis, thus avoiding further reduction in chromosome number.

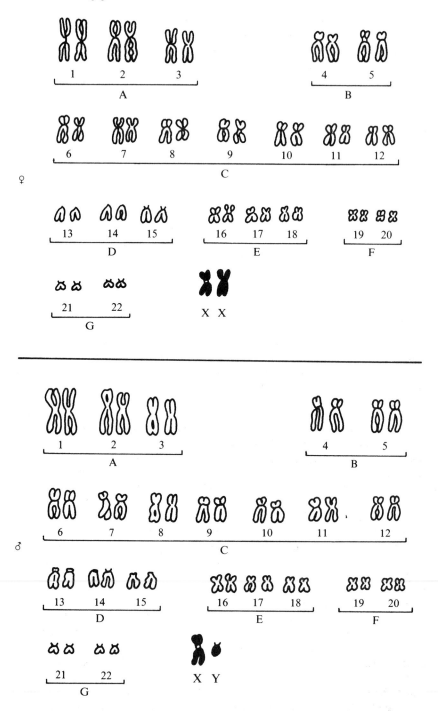

Fig. 8.2 Karyotypes in man showing the difference in chromosome complements between the sexes. The X and the Y are the sex chromosomes and pairs 1–22 the autosomes.

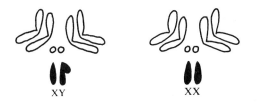

Fig. 8.3 Diagrammatic representation of the chromosome complements in male and female *Drosophila melanogaster* showing the three pairs of autosomes (in outline) and the X and Y sex chromosomes.

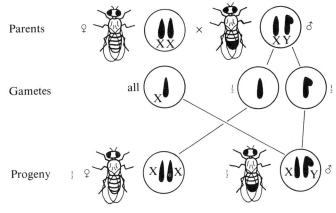

Fig. 8.4

Sex-linked genes

Sex chromosomes also carry genes which are not directly involved with the determination of sex. The characters which these genes control are associated in their inheritance with the character of sex. For this reason *genes which are located in the sex chromosomes, and which are linked to the sex-determining genes*, are called **sex-linked genes**.

The inheritance of sex-linked genes is different from that for autosomal genes, because in the heterogametic sex the sex chromosomes are not completely homologous with one another. In the XX/XY system, with which we are mainly concerned, there are certain genes in the differential part of the X chromosome which have no allelic partners in the Y and vice versa. A general scheme for the structure and genetic organisation of the XY sex chromosome pair is shown in Fig. 8.5.

Genes in the homologous pairing segment, in which there may be crossing over between the X and Y, show **partial sex linkage**. They have normal diploid inheritance like those in the autosomes. Their presence in the sex chromosomes is detected because they show linkage to the genes in the differential segment, which are the ones with the distinctive patterns of sex-linked inheritance.

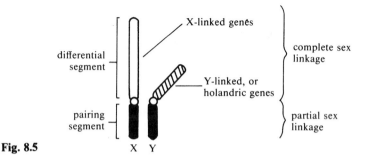

Fig. 8.5

Genes in the differential segments have no crossing over between them and are said to show **complete sex linkage**. As it happens, the Y chromosome in most species is largely devoid of genes, as in *Drosophila* and man, and the pairing segment is often quite small. In talking about sex-linked inheritance, therefore, we are mainly concerned with the genes in the differential segment of the X chromosome.

Inheritance of sex-linked genes

What is the pattern of inheritance of sex-linked (i.e. X-linked) genes, and how do we detect such genes?

Eye colour in *Drosophila*

The first evidence of sex linkage came in 1910 from an experiment by Morgan and his co-workers using *Drosophila*. Amongst his cultures of normal red-eyed flies Morgan found a white-eyed male. He isolated this male and crossed it to its red-eyed sisters. He found that all the F_1 had red eyes, showing that white-eyed was recessive. When the F_1 were mated amongst themselves he found red-eyed and white-eyed flies in a 3 : 1 ratio in the F_2, *but all the white-eyed flies were male.*

Morgan was puzzled by the absence of white-eyed females in the F_2. He found that if he took the red-eyed female F_1 and backcrossed them to the white-eyed male parent, the testcross progeny comprised white-eyed and red-eyed males and females in a 1 : 1 ratio. This showed that it *was* possible for a female to have white eyes and also that the female F_1 progeny *were* heterozygous in genotype. How then can we explain the absence of white-eyed females in the original F_2? The explanation arises because the gene concerned is located in the X chromosome, i.e. it is X-linked. We can represent the cross using the notation X^w, to indicate the location of the gene on the X, and Y to indicate the Y chromosome (Fig. 8.6).

Once the gene is located on the X we can see how the results are obtained. The female in *Drosophila* is the homogametic sex, and so for the gene concerned she can be:

$$X^{w^+}X^{w^+} \quad \text{homozygous, red-eyed}$$
$$X^{w^+}X^{w} \quad \text{heterozygous, red-eyed}$$
$$X^{w}X^{w} \quad \text{homozygous, white-eyed.}$$

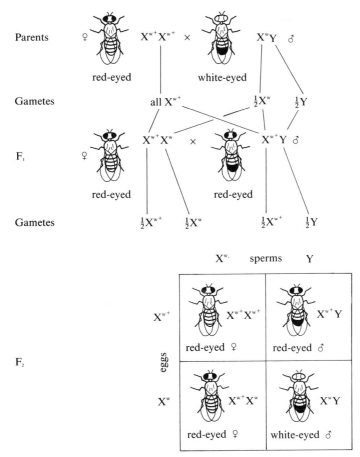

Fig. 8.6   3 : 1 red : white eyes.

The male, however, only carries one X chromosome and has no corresponding gene in the Y. As a consequence, whatever is carried on the X chromosome is obvious in the phenotype of the male, because it is present singly, and there is no other member of the chromosome complement whose genotype can mask it. *Genes present only once in the genotype, and not in the form of pairs*, are said to be **hemizygous**. The male can therefore be:

$$X^{w^+}Y \quad \text{hemizygous, red-eyed}$$
$$X^wY \quad \text{hemizygous, white-eyed.}$$

The female parent was homozygous red-eyed and produced only one kind of gamete, X^{w^+}. The male was hemizygous white-eyed and produced two kinds of gamete X^w and Y. In the F_1 both male and female are red-eyed but the males are hemizygous ($X^{w^+}Y$) and the females heterozygous ($X^{w^+}X^w$). In the F_2, half of the male progeny received X^{w^+} from the female (and Y from the male) and half receive X^w. Because they are hemizygous, the genotype is apparent in the phenotype and half the males are red-eyed and half white-eyed. The female F_2 progeny receive X^{w^+} from the male and either X^{w^+} or X^w from the female. They are therefore all red-eyed.

In this experiment sex linkage was indicated because of the difference between males and females in the F_2. Females were all red-eyed and males $\frac{1}{2}$ red and $\frac{1}{2}$ white. In all other monohybrid crosses we have examined so far, which were not concerned with sex-linked genes, the pattern of inheritance was independent of sex.

The second indication of sex linkage comes when reciprocal crosses are made. When the female is now used as the white-eyed parent, and the male as red-eyed, there is a different outcome to that shown in Fig. 8.6. Both red-eyed and white-eyed flies are found in the F_1 and there is now a 1:1 ratio of red:white in the F_2 (Fig. 8.7).

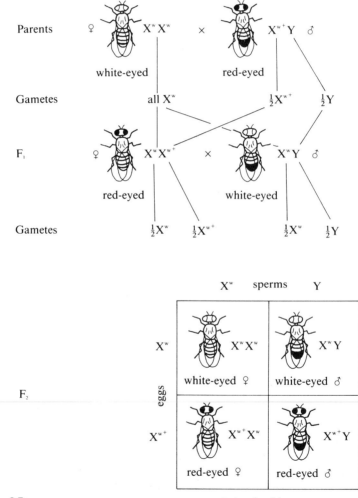

Fig. 8.7   1:1 red:white eyes.

We notice that in this second cross the females in the F_1 resemble the eye-colour character of their fathers, and the males that of their mothers. This pattern is referred to as **criss-cross inheritance** and is another indication that a character is controlled by a sex-linked gene.

Sex linkage in man

A number of human characters appear much more frequently in males than they do in females, and show a pattern of inheritance which is consistent with sex linkage. The best known ones are haemophilia and colour blindness.

Haemophilia   Haemophilia is a disease of the blood. The commonest form is caused by a single recessive gene. Its main feature is a reduced ability of the blood to clot so that haemophiliacs are susceptible to protracted bleeding, even from trivial wounds. Haemophilia has been carefully pedigreed in the royal family of Queen Victoria. The lineage is shown in Fig. 8.9 overleaf. The fact that only male members of the family showed the disease is indicative that this is a sex-linked character. In humans, we recall, the female is the homogametic sex and so can be:

$$X^H X^H \qquad \text{normal, homozygous}$$
$$X^H X^h \qquad \text{normal, heterozygous 'carrier'}$$
$$X^h X^h \qquad \text{haemophiliac, homozygous.}$$

The male, on the other hand, is heterogametic and does not carry the haemophilia locus on the Y chromosome. Males can therefore be:

$$X^H Y \qquad \text{normal, hemizygous}$$
$$X^h Y \qquad \text{haemophiliac, hemizygous}$$

Since one of Queen Victoria's sons was a haemophiliac and Albert of Saxe-Coburg-Gotha (her husband) was normal, it is clear that Queen Victoria must have been a carrier. We can represent the marriage as shown in Fig. 8.8.

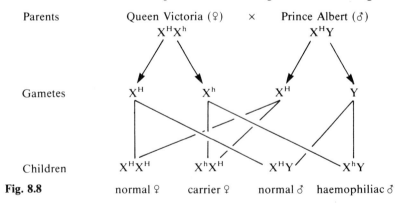

Parents   Queen Victoria (♀)   ×   Prince Albert (♂)
                        $X^H X^h$ $X^H Y$

Gametes X^H X^h X^H Y

Children $X^H X^H$ $X^h X^H$ $X^H Y$ $X^h Y$

Fig. 8.8   normal ♀   carrier ♀   normal ♂   haemophiliac ♂

Queen Victoria produces two types of gametes, half of which carry **H** on the X chromosome and half carry **h**. On average, therefore, we would expect that half her sons would receive X^H (and be normal $X^H Y$), and half would receive X^h (and be haemophiliac $X^h Y$). Her daughters receive an X with the normal H allele from Prince Albert, and either X^H or X^h from Queen Victoria. They will therefore be either carriers, or homozygous normals. Some of her daughters were clearly carriers, since haemophiliac males appear in their children and in total three generations were affected by the disease. Female haemophiliacs are only rarely found because they depend on marriages between haemophiliac males and carrier (or haemophiliac) females, which will not occur very often.

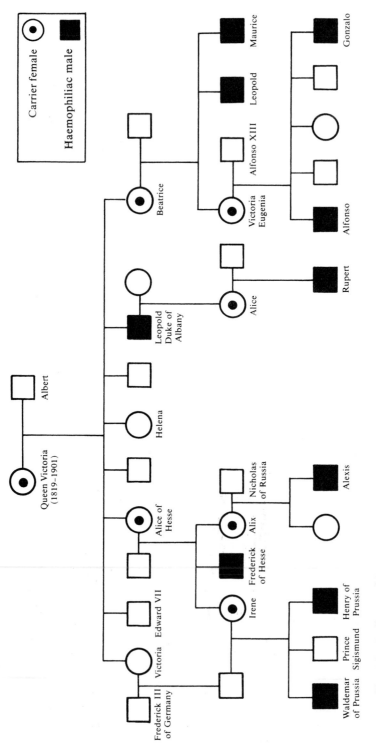

Fig. 8.9 Pedigree of haemophilia in the royal families of Europe (excluding many individuals from later generations).

Colour blindness In humans, one form of colour blindness results from a recessive gene carried on the X chromosome. Now consider the following question.

A colour-blind boy had parents with normal colour vision.
(*a*) What is the genotype of (i) the mother?
                                (ii) the father?

Answer: Since both parents are normal, the mother must have been a carrier. Using **C** for the normal allele of the gene and **c** for the allele causing colour blindness we can represent the mother as X^CX^c. The father was also normal and must have been X^CY.

(*b*) If the boy's brother, who has normal colour vision, marries a girl who is heterozygous for colour vision, what would be the possible phenotypes and genotypes of their children?

Answer: The boy's brother is normal and so must be X^CY, like his father. The phenotypes and genotypes of their children would be as in Fig. 8.10.

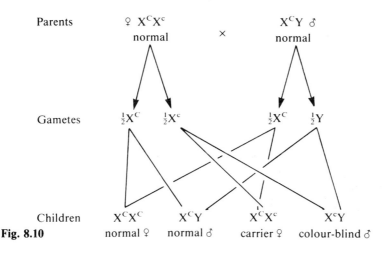

Fig. 8.10

Other X-linked genes in man include 'Christmas disease': which was named after the patient in whom it was first described. This is another type of haemophilia caused by reduction in the amount of plasma thromboplastin component (PTC), and is called haemophilia *B* to distinguish it from haemophilia *A* (the type described in Queen Victoria's family), which results from reduction in the amount of anti-haemophiliac factor (AHF).

Y-linked genes in man There are very few known examples of Y-borne genes in man. Some pedigrees suggesting Y linkage have been published, including one for a defect described as porcupine skin and one for webbed toes, but fuller examination raised scepticism as to their validity. A Y-linked gene can only ever be found in males and a possible example is that of the character hairy ears, which has been traced over seven generations in one particular pedigree and was only shown by the male members.

Sex linkage in birds

As we have already mentioned, in birds, butterflies and moths, the female is the heterogametic sex and the male is homogametic. The same principle of inheritance applies as the one we have described for sex linkage in man and in *Drosophila*, but the pattern of inheritance is reversed in the sexes. For an example consider the following problem:

Light Sussex fowls have mostly white plumage, and the Rhode Island Red breed have mostly red. The character 'white-feathered' (**R**) is dominant to the character 'red' (**r**). Explain why on crossing Rhode Island cockerels with Light Sussex hens all male offspring have white plumage and all females are red, as shown in Fig. 8.11.

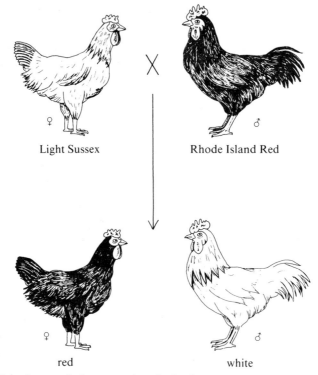

Light Sussex   Rhode Island Red

red   white

Fig. 8.11 Inheritance of plumage colour in fowls.

The difference between males and females in the progeny, and the criss-cross inheritance, indicates that this is a sex-linked character. The female parent was white-feathered and, since the female is the heterogametic sex in birds, she must have been $X^R Y$.

The male was red in colour, and since he is the homogametic sex, and red is recessive, he must have been $X^r X^r$. We can now explain the cross (Fig. 8.12).

The Light Sussex female produces two kinds of gametes, X^R and Y. The Rhode Island Red male produces only one type, X^r. Female progeny will receive Y from their mother and X^r from their father and will be red. Male progeny will receive X^r from their father and X^R from their mother and will be white.

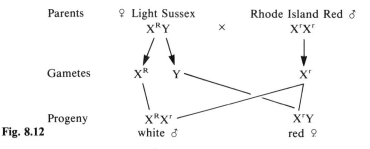

Parents   ♀ Light Sussex   Rhode Island Red ♂
                   $X^R Y$ × $X^r X^r$

Gametes X^R Y X^r

Progeny $X^R X^r$ $X^r Y$
Fig. 8.12   white ♂   red ♀

Box 8.1   Mechanisms of sex determination

Sex is a complex developmental character and it seems that there are many genes involved in the differentiation into male and female forms. Most of these genes are located in the autosomal chromosomes. The main function of the sex chromosomes is to act as a 'switching' device and to direct the developmental process along one of the two alternative pathways. The way in which the sex chromosome switch works is quite different for man and for *Drosophila*, even though both have the same XX/XY system.

In *Drosophila* the mechanism is based on a balance between the number of Xs and the number of autosomal sets. The Y is needed for the sperms to function properly but it is not essential for determining maleness. A fly that has two Xs (XX) and two sets of autosomes (AA), i.e. where the ratio of X/A = 1.0, is female. A fly with a single X and an X/A ratio of 0.5 is male. An autosome set in *Drosophila* comprises chromosomes 2, 3 and 4, chromosome 1 being the X. A normal diploid has two autosome sets which we represent as AA.

The idea that sex determination in *Drosophila* depends upon this *balance* of Xs to autosomes came about from the discovery of a **gynandromorph** (*a sexual mosaic typically male in certain parts of the body and female in others*). A fly was found which was half male and half female (Fig. 8.13). The male part was made up of cells containing a single X, and the female part of cells with two Xs. The zygote which produced this fly was XX and therefore destined to be female, but at the first cell division of the zygote an X chromosome was lost from the nucleus of one of the daughter cells and this cell gave the lineage for the male half of the fly. The point about this gynandromorph was that the typically male part of the body did not possess a Y chromosome.

In man the mechanism is different. Gynandromorphs are not possible, because sexual differentiation is influenced by hormones circulating in the blood system, and is not determined on an individual cell basis as in the fruit fly. Chromosome mutants do arise in the population, however, at mutation frequency (Chapter 16) and these can be used to obtain information about the role of the X and Y chromosomes in the switching mechanism.

It turns out that any individual which has one, or more, Y chromosomes is a male (i.e. has testes), irrespective of the number of Xs; and any individual with one or more Xs (but without any Ys) is female, with ovaries. Persons with abnormal numbers of sex chromosomes are often sterile and have a complex syndrome of other phenotypic abnormalities as well—but they can generally be classified as either male or female. From a study of such individuals, it is apparent that in man the presence of the Y chromosome determines maleness and its absence femaleness. In addition to normal females with a chromosome complement of 22 pairs of autosomes plus XX, mutant types have been found with only one, or three, four and five X chromosomes. Likewise males have been identified, in addition to normal ones with 22 pairs of autosomes plus XY, having 22 + XYY, 22 + XXY, 22 + XXYY, 22 + XXXY and 22 + XXXXY.

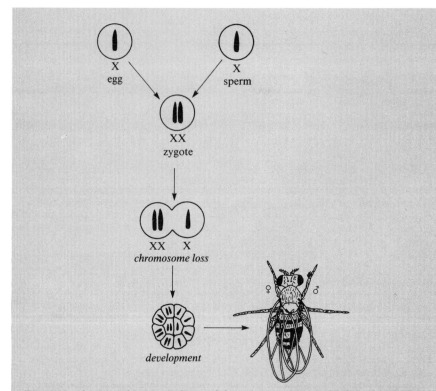

Fig. 8.13 A gynandromorph *Drosophila*. When a zygote which is destined to become a female fly (XX) loses an X chromosome from one of the two daughter nuclei at the first mitosis, it can result in a fly which is half male and half female. The sex chromosome constitution of the male half is XO. (Autosomes not shown.)

In man, as in *Drosophila*, the sex of an individual is determined at fertilisation by the male gamete. X-carrying sperms give females and Y-carrying sperms males. It is now possible in man to separate out the two kinds of sperms from a sample of semen on the basis of differences in the electrical charges on the nucleus due to slight differences in structure between the X and Y chromosomes. Using artificial insemination the sex of a child can be predetermined.

X chromosome inactivation

The difference in structure between the sex chromosomes of males and females has consequences for the activity and expression of sex-linked genes as well as for their inheritance. In the XX/XY system of *Drosophila*, and of many mammals, the Y chromosome has few genes. Males are hemizygous for loci in the differential segment of the X chromosome while females have two Xs and two copies of each of their X-linked genes. This difference poses a problem in terms of the imbalance in X-linked genes between the two sexes.

In *Drosophila* the imbalance is dealt with by a regulatory process in which the single X in males 'works' twice as hard, transcribing messenger RNA (Chapter 14) at twice the rate of each of the two Xs in the females.

Mammals have a different regulatory system of dosage compensation whereby one of the two Xs in the female is 'switched off' in early development and rendered genetically inactive. In 1949, M. L. Barr discovered that the nuclei of cells from the brains of female cats contained a dark-staining body which became known as the **Barr body**. Interest was aroused in these Barr bodies when it was found that their presence or absence was correlated with the number of X chromosomes of an individual. In humans, for example, the number of Barr bodies is always one less than the number of Xs (Fig. 8.14).

Sex chromosomes of the individual	No. of Barr bodies in somatic cell nuclei
♂ { XY	0
XYY	0
XXY	1
♀ { XX	1
XXX	2

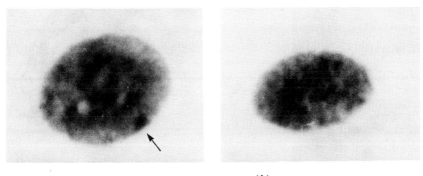

(a)   (b)

Fig. 8.14   Nuclei of epithelial cells taken from the inside of the cheek of a normal human female (a) and male (b). The female nucleus shows the prominent Barr body (arrowed) lying adjacent to the nuclear membrane. (Photographs by courtesy of Lindsay Hague.)

At a later date, Lyon and Russell independently put forward the theory that the Barr bodies are formed as a result of the inactivation of all but one of the Xs by a process of **heterochromatinisation**. Heterochromatin in this context means a tightly coiled-up form of the interphase chromatin of the X or Xs concerned, such that the DNA molecules within them are unavailable for transcription by messenger RNA, and there is thus no expression of their genes. This idea subsequently became known as the **Lyon hypothesis** and it is the way in which the balance between the sexes is achieved for their X-linked genes. A normal female mammal with XX has one Barr body, whilst a normal male, XY, has none. Dosage compensation therefore takes place by random inactivation of one of the two Xs in each cell of the female somatic tissues. In individuals with an abnormal number of Xs the inactivation affects all except one, even in males (e.g. XXY in man).

The genetic effect of X chromosome inactivation can be seen in the phenotype for sex-linked genes which are expressed at the level of individual cells. An example is coat colour in cats. One well known gene gives black fur when dominant (**B**) and 'ginger' (sometimes called 'yellow') when recessive (**b**). Males are hemizygous and may be either black (X^BY) or ginger (X^bY). Females have three possible genotypes: X^BX^B, which is black, X^bX^b, which is ginger and the heterozygote, X^BX^b, which is a mosaic of both black and ginger patches—except on the belly which is white. This 'tortoise-shell' appearance of the female results from random inactivation of one of the two Xs early in development, after which the cells multiply without further change to give patches of either black or ginger fur depending upon which allele is present in the active X. Tortoise-shell coat colour patterns are also found in certain males which carry a chromosome mutation, XXY. These occur with a frequency of about 1 in 3000 among the cat population and the affected individuals are sterile.

Mosaicism for X-linked genes is also found in human females. An example is the mutation which causes deficiency of the blood cell enzyme glucose-6-phosphate dehydrogenase (G6PD). In heterozygotes, X^GX^g, half the population of blood cells in the body carry the enzyme and half are deficient. The phenotype is normal.

Summary

Sexual differentiation into male and female forms within species is mainly determined by particular chromosomes within the complement, called sex chromosomes. These differ in their form, and gene content, and in the ways in which they bring about the determination of sex.

Genes located in the sex chromosomes show modified patterns of inheritance known as sex linkage, because the characters they determine are associated in their inheritance with sex.

The sex chromosome difference causes an imbalance in gene content between males and females. Regulatory mechanisms of gene action have evolved which compensate for the difference in gene dosage between the sexes.

Questions and problems

1 In *Drosophila*, white eye (**w**) is recessive to the normal red eye (**W**), is sex-linked and is situated on the X chromosome.
   (a) What is meant by 'recessive'?
   (b) What is the significance of the symbols **w** and **W**, for white eye and red eye respectively?
   (c) What is the meaning of 'sex linkage'?
   (d) What symbol is used for the partner chromosome of X:
      (i) in the male fly
      (ii) in the female fly?
   (e) Show by simple concise diagrams:
      (i) the results of crossing a white-eyed female with a red-eyed male.
      (ii) the results of crossing an F_1 male with an F_1 female derived from the cross in (e)(i).
                                                      (O.L.E. Zool., 1981)

2 Briefly explain what you understand by *three* of the following:

   sex linkage; incomplete dominance; probability; wild type; pure line.

   In the fruit fly *Drosophila melanogaster*, white eye colour is recessive to vermilion and vestigial wings to normal wings. These genes are located on the X chromosome. In *Drosophila* the male is the heterogametic sex. If a homozygous white-eyed, normal-winged female is crossed with a vermilion-eyed, vestigial-winged male, what will be the phenotypes and their proportions in the offspring of (a) the F_1 generation, (b) the F_2 generation, when F_1 flies are allowed to interbreed, and (c) the offspring of a cross between the F_1 flies and each of the parental types? You should show clearly each stage of your working in arriving at your conclusions.

(O.L.E. Biol., 1981)

3 (a) Distinguish between asexual and sexual reproduction. Explain the significance of the differences between these two types of reproduction (see Chapter 18).

 (b) In *Drosophila*, normal wings are produced by the dominant gene **M** and miniature wings by the recessive allele **m**. In an experiment, pure-breeding female *Drosophila* with normal wings were crossed with male flies having miniature wings. The resulting F_1 generation consisted of male and female flies with normal wings.

 (c) When some of these flies were allowed to interbreed, the following progeny were obtained:

>   105 females, with normal wings
>   52 males, with normal wings
>   46 males, with miniature wings.

   Give a reasoned explanation of these results.

   Describe the procedure which would be required in order to obtain female *Drosophila* with miniature wings, using flies from the F_2 generation of the previous experiment. What other phenotypes might also be produced?

(L. Biol., 1980)

4 Define the terms 'total linkage', 'partial linkage' and 'sex linkage'. How does linkage cause variations in Mendel's laws?

   Show how a sex-linked character (e.g. haemophilia in man) is transmitted from one generation to the next in named organisms. What importance has linkage in the evolutionary process? *(S.U.J.B. Biol., 1982)*

5 A man whose father suffers from haemophilia, and whose mother is normal and not a carrier, cannot suffer from this inherited disease. Explain this, using symbols to represent genes. *(S.U.J.B. Biol., 1980)*

6 In the following human pedigree (Fig. 8.15), circles indicate females, squares males. Filled-in circles and squares indicate red–green colour-blind individuals; open circles and squares indicate normal vision.

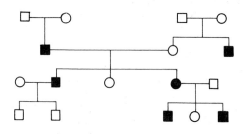

Fig. 8.15

State, as far as possible, the genotypes of each individual in the pedigree. What conclusions can you draw as to the inheritance of colour blindness?

7 Evidence has accumulated that a hospital may have mixed up two girls who were born within a short time of each other. Legal proceedings have been started by one family and scientific evidence is called for to settle the case. The main blood groups were identical in the parents and in the supposed daughter. However, the mother and the supposed daughter were colour-blind while the father was not. The gene for colour blindness is recessive and is present on the X but not on the Y chromosome.

(a) What are the genotypes of the parents?

(b) By means of a diagram, indicate the possible genotypes of offspring of these parents.

(c) What are the phenotypes of these possible offspring?

(d) What is the genotype of the supposed daughter?

(e) Can the supposed daughter belong to these parents? Yes or no?

8 'Black and tan' is a character in mice in which the development of black fur is restricted to the top side of the body. A pure-breeding black and tan female with long ears was crossed with a pure-breeding black, short-eared male. The offspring had black and tan fur and long ears. These F_1 mice were interbred and the following progeny obtained.

	Black & tan, long ears	Black, long ears	Black & tan, short ears	Black, short ears
Females	16	7	6	2
Males	20	5	4	2
Total	36	12	10	4

(a) Is there any evidence from the data to suggest that the characters of fur colour and ear length in mice are sex-linked?

(b) Explain your answer given in part (a) above.

(c) What else can be deduced from the data about the nature and location of the genes which control fur colour and ear length?

(d) In the space below, construct a diagram to explain the probable result of a cross between an F_1 individual and its short-eared parent.   (L. Biol., 1980)

9 By what mechanisms may sex be determined in animals? In Drosophila, where the male is heterogametic, a gene for eye colour is located on the X chromosome. The wild-type eye is red but there is a recessive mutant allele, w, for white eye colour. What are the genotypes of white-eyed males and females?

With the help of diagrams show what would be the possible genotypes and phenotypes of the F_2 progeny of crosses between:

(a) a white-eyed male and a red-eyed female;

(b) a red-eyed male and a white-eyed female.   (O.L.E. Zool., 1982)

10 A pair of alleles, B and b, appear to be sex-linked in cats. In females, homozygous BB gives a black coat colour, homozygous bb gives a ginger coat colour and heterozygous Bb gives the familiar tortoise-shell with a patchwork of well-defined areas of ginger and black.

(a) (i) How is sex determined in animals and what is meant by the term 'sex-linked'?

(ii) Give another example of a sex-linked character.

(b) A cross was made between a tortoise-shell female and a ginger male cat. Explain, by use of a diagram, the genotypes of the parents and F_1 progeny of this cross.

(c) What are the possible phenotypes and sexes of the F_1 progeny?

(d) A tortoise-shell cat had a litter of 8 kittens: 1 ginger male, 2 black males, 3 tortoise-shell females and 2 black females. Assuming that there was a single father for the litter, what was his coat colour?

(e) How would you explain the patchwork coat of the tortoise-shell cat?

(C. Biol., 1982)

11 (*a*) How is sex genetically determined in birds and humans?

   (*b*) (i) A woman has a haemophiliac son and three normal sons. What is her genotype and that of her husband with respect to this gene? Explain your answer.

      (ii) Could she have a haemophiliac daughter? Explain your answer giving your reasons.

   (*c*) A population of human beings will contain more colour-blind individuals than haemophiliacs although the genes are transmitted in the same way. Explain this difference in frequency.   *(L. Biol., 1983)*

9
Modified genetic ratios

Crosses between pure-breeding individuals which differ in one or two pairs of genes do not always give the classical $3:1$ and $9:3:3:1$ Mendelian ratios in the F_2. One reason is that the genes concerned may be linked (Chapter 7). Another reason is sex linkage (Chapter 8). There is also a further complication arising from the fact that genetic ratios depend upon the way in which genes are *expressed* in the phenotype, as well as upon their mode of *transmission*. So far we have dealt only with simple examples of gene action. For a single character, determined by one gene, one of the alleles has always shown complete dominance over its partner in the heterozygote. Where two independently-segregating genes (**A** and **B**) have been involved, they have been acting separately from one another and controlling two different characters. Gene action is not always this straightforward.

Alleles of a single gene may **interact** together and give rise to F_1 phenotypes which are dissimilar to both of the two parents and to F_2s which have a modified ratio of phenotypes. Likewise two independently-inherited genes may both affect the same single character and this again will lead to modifications of the familiar $9:3:3:1$ ratio in the F_2.

In this chapter we will deal with some of the well known examples of **modified genetic ratios**, due to gene interaction, and also briefly mention the pattern of inheritance displayed by genes located outside of the nucleus, in cell organelles.

Interaction between alleles

Incomplete dominance

In cases of **complete** (or **simple**) **dominance** *the heterozygote* (**Aa**) *has the same phenotype as the dominant homozygote* (**AA**). This is the relationship between a pair of alleles that we have been concerned with in the previous chapters— because it provides the simplest situation for study. With **incomplete dominance** (**semi-dominance**, **partial dominance**) *the heterozygote exhibits a phenotype which is intermediate between the two homozygous forms*. An example is found in the four-o'clock plant, *Mirabilis jalapa*. When pure-breeding red-flowered four-o'clocks are crossed to pure-breeding white-flowered ones, all the F_1 have pink flowers. Selfing the F_1 gives a ratio of 1 red:2 pink:1 white among the F_2. The cross may be represented as shown in Fig. 9.1.

The genotypic ratio is exactly what would be expected for a monohybrid cross, but the phenotypic ratio is modified to $1:2:1$, and is the same as the genotypic ratio, because incomplete dominance results in the heterozygotes being intermediate between the two homozygous forms.

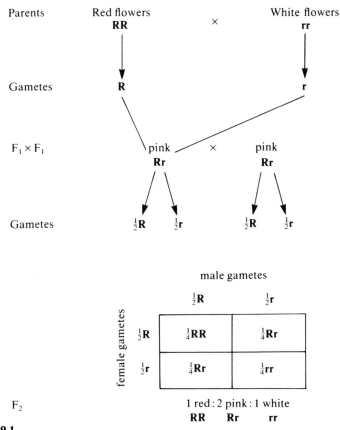

Fig. 9.1

Another example of incomplete dominance occurs in domestic fowl. In pure-breeding Andalusians, black feather colour is due to the formation of the pigment melanin. A pure white strain was found which was lacking in melanin. The F_1 between the two strains has an intermediate colour called slate blue, which is due to partial development of melanin. Crossing F_1s gives a $1:2:1$ ratio of black:blue:white in the F_2 (Fig. 9.2).

It was most probably these kinds of observations which led to the earlier idea of 'blending inheritance', which we now know to be incorrect (Chapter 5). The point is that although the F_1 are intermediate in character the parental types segregate out again in the F_2. This shows that the alleles themselves have not 'blended' together in the F_1, but have remained separate from one another as discrete units of inheritance. It is their *interaction*, at the level of gene action and expression, which gives rise to the intermediate phenotype.

Co-dominance

In **co-dominance** *both alleles are expressed in the phenotype and the heterozygote has the characters of both parents.* The MN blood group system in man provides us with an excellent example. Antigenic proteins on the surface of the red blood cells give individuals their blood group phenotype. There are three

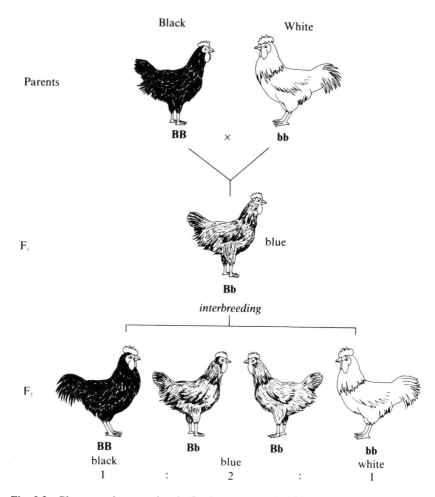

Parents

F₁

interbreeding

F₂

Fig. 9.2   Plumage pigmentation in fowl—an example of incomplete dominance.

blood groups in this system, M, N and MN, determined by two co-dominant alleles of a single gene. People who are homozygous **MM** have blood group M and produce only M antigens on their red cells; **NN** homozygotes are group N, with N antigens; and heterozygotes, **MN**, are group MN and produce both M and N antigens. In a cross **MM × NN** the F₁ (**MN**) have the characters of both parents, i.e. both kinds of antigens, and in the F₂ there is a modified 1 : 2 : 1 phenotype ratio of 1 M : 2 MN : 1 N due to the co-dominance. The blood group genetics of man is discussed in more detail in Chapter 10.

Lethal alleles

Lethal alleles are not actually a form of gene interaction, but since they also lead to modified F₂ ratios they can be dealt with conveniently in this section.

In 1905 the French geneticist Lucien Cuénot studied a strain of mice which had yellow fur instead of the normal grey (agouti). On crossing yellow with pure-breeding grey he obtained a 1:1 ratio of yellow:grey among the progeny, indicating that yellow mice were heterozygous (**Yy**) and that yellow was dominant. When two yellows were mated together the progeny were in the ratio of 2 yellow:1 grey, which was a puzzling deviation from the expected 3:1. The clue to this deviation emerged when it was realised that pure-breeding homozygous yellows (**YY**) were never found, and that all yellow mice were heterozygotes. Cuénot concluded that the gene for yellow must be lethal in the homozygous form. This was confirmed by finding aborted foetuses in **Yy** females. The modified 2:1 ratio is therefore explained (Fig. 9.3).

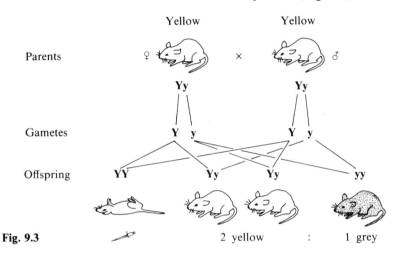

Fig. 9.3

An interesting aspect of gene action in this example is the way in which allele **Y** is dominant in its effect on coat colour, because the heterozygotes (**Yy**) are yellow, and recessive in terms of its effect of lethality—since it only results in death when present in the homozygous form. We therefore say that **Y** is a dominant visible and a recessive lethal gene.

A common form of lethal genes in plants are those which cause a lack of chlorophyll. These are usually recessive mutations and death occurs in the seedling stage because homozygous recessives are unable to engage in photosynthesis.

Interaction between different genes

Examples of modified ratios due to interaction between different genes are given below and also in Box. 9.1. They all refer to effects due to two pairs of unlinked, independent, genes with complete dominance between alleles.

Duplicate genes (15:1)

Duplicate genes *are two different genes which affect the same character in the same way*, e.g. seed capsule shape in the shepherd's purse (*Capsella bursa-pastoris*). When pure-breeding plants with round capsules are crossed to those with narrow capsules the F$_1$ all have round capsules. In the F$_2$ there is a phenotypic ratio of 15:1 for plants with round or narrow capsules. The interpretation of this cross is that the character of capsule shape is controlled by two genes with the same effect. A dominant allele of either gene (A_1 or A_2) is sufficient to give round capsules, and narrow capsules are found only in individuals which are homozygous recessive for both genes. The fact that it is a 15:1, and not a 3:1 ratio indicates that two different independently-inherited genes are controlling the character, and not just a single gene (Fig. 9.4.).

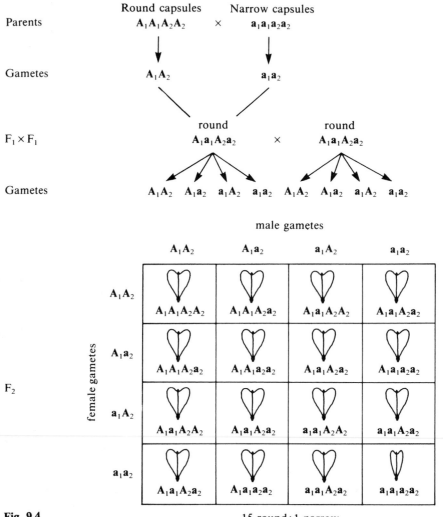

Fig. 9.4   15 round:1 narrow

Complementary genes (9:7)

Complementary genes *interact together to produce an effect which is different from that given by either of them separately.*

Bateson and Punnett crossed two white-flowered sweet peas and found that their F_1 had purple flowers! When the F_1s were crossed together they produced

F_2s with purple and white flowers in a ratio of 9:7. The explanation for this result (which is a common form of gene interaction) is that the character of flower colour is controlled by two different genes which are **complementary** to one another in their mode of action. One of the genes (designated as **C**) controls the production of a colourless pigment precursor, and another one (**P**) controls the conversion of the pigment precursor into its purple form. A dominant allele of both of the genes has to be present to give purple flowers. If either of the dominant alleles is absent the flowers are white, either because there is no pigment precursor produced, or else it is produced but not converted into the purple form. If both dominant alleles are lacking then there is neither production of the precursor nor conversion to the pigment, and the flowers are again white. Clearly the white-flowered parent strains must have been of different genotypes: each was homozygous dominant for a different gene since the F_1 progeny were purple and received a dominant allele from both parents. The parents were therefore **CCpp** and **ccPP**, and white-flowered for different reasons. The F_1 progeny contained both dominant alleles and were therefore purple. In the F_2 only two phenotypic classes are observed because of the complementary way in which the two genes interact together to determine the phenotypes (Fig. 9.5).

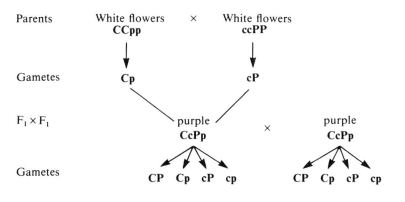

Parents	White flowers **CCpp**	×	White flowers **ccPP**

Gametes	**Cp**		**cP**

$F_1 \times F_1$	purple **CcPp**	×	purple **CcPp**

Gametes	**CP Cp cP cp**		**CP Cp cP cp**

male gametes

		CP	Cp	cP	cp
female gametes	**CP**	**CCPP** purple	**CCPp** purple	**CcPP** purple	**CcPp** purple
	Cp	**CCPp** purple	**CCpp** white	**CcPp** purple	**Ccpp** white
	cP	**CcPP** purple	**CcPp** purple	**ccPP** white	**ccPp** white
	cp	**CcPp** purple	**Ccpp** white	**ccPp** white	**ccpp** white

F_2

Fig. 9.5   9 purple : 7 white

If the initial cross had been made between a pure-breeding purple parent (**CCPP**) and a pure-breeding white one (**ccpp**), the same heterozygous F_1, and the same $9:7$ ratio in the F_2, would have been obtained. It just so happened that Bateson and Punnett made their original discovery after crossing together two white-flowered plants.

Comb form in the fowl

Bateson and Punnett also worked out the rather unusual gene interactions that are involved in determining the comb type in poultry. There are a number of different forms of the comb, two of which are called 'pea' and 'rose'. Both of these forms breed true, but when crossed together the resultant F_1 hybrid has an entirely different form called 'walnut'. When walnut fowls interbreed, the next generation has four comb types: walnut, pea, rose and a new type known as 'single'. The four phenotypes occur in a ratio of $9:3:3:1$ (Fig. 9.6).

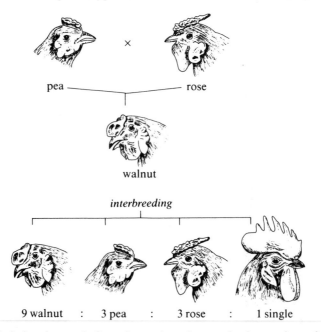

Fig. 9.6 Variation in comb form in poultry due to the interaction of two genes controlling the same character. (After Punnett.)

Bateson and Punnett explained this cross on the basis of the interaction of two independent genes which were called **P** (for 'pea') and **R** (for 'rose'). Pea and rose are both dominant over single. A genotype which has at least one dominant allele of gene **P** has a pea comb, and one with at least one dominant allele of gene **R** has a rose comb. A genotype without any dominant alleles has a single comb, which results from the interaction between the two homozygous recessive pairs of alleles. When the dominant alleles of both **P** and **R** are present together they interact to give the walnut phenotype. Pure-breeding parents with pea combs are therefore of genotype **PPrr** and those with rose combs are **ppRR**. The F_1 are walnut because they have the dominant alleles of both genes (**PpRr**).

The important clue to working out the genetics of this cross, and of the other examples given in this section, is the way in which the different phenotypic classes are seen to be in multiples of 1/16ths. This tells us that the F_2 must have come about by the crossing of F_1s which were heterozygous for two pairs of independently-segregating genes.

It is important to realise that this example, while it is a 9:3:3:1 ratio, differs from the normal dihybrid cross in that *there is only one character involved*— comb form. The non-parental forms which turn up in the F_2 are not recombinants as such, since they do not include new combinations of characters already present in the parents. They are completely *novel* forms of the one character due to interactions between its two controlling genes (Fig. 9.7).

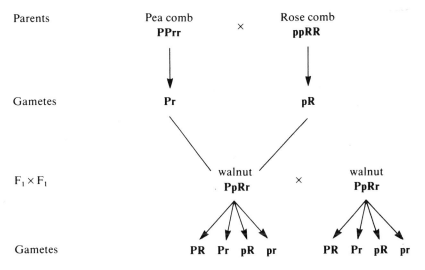

		PR	Pr	pR	pr
		PPRR walnut	PPRr walnut	PpRR walnut	PpRr walnut
	PR				
	Pr	PPRr walnut	PPrr pea	PpRr walnut	Pprr pea
	pR	PpRR walnut	PpRr walnut	ppRR rose	ppRr rose
	pr	PpRr walnut	Pprr pea	ppRr rose	pprr single

Fig. 9.7   9 walnut : 3 pea : 3 rose : 1 single

Box 9.1   Epistasis

In **epistasis** (='standing upon') *one gene hides the expression of another one.*
Epistasis is involved in many kinds of gene interactions, including some of those
we have already discussed. In the shepherd's purse, for instance, a dominant
allele of either of the two duplicate genes is epistatic to the recessive homozygote
of the other one. In genotypes $A_1A_1a_2a_2$, or $A_1a_1a_2a_2$, A_1 is epistatic and hides
the narrow capsule phenotype controlled by a_2a_2. Likewise in $a_1a_1A_2A_2$, or
$a_1a_1A_2a_2$, the A_2 allele is epistatic over a_1a_1.

In the sweet pea, with its two complementary genes determining flower colour,
we may say that either recessive homozygote is epistatic to the expression of the
other gene, because a dominant allele of both genes is required to produce purple
flowers.

Another well known example of epistatic gene action is that of coat colour in
mice. Wild mice have agouti fur. The individual hairs are black with a yellow
band, and this combination of colours produces the typical mouse agouti
phenotype. Non-agouti mice have black coats, composed of uniformly black
hairs. The difference is due to a single gene with two alleles, **A** and **a**. Agouti
(**AA**, **Aa**) is dominant over black (**aa**). A second independently-inherited gene
is required for synthesis of the hair pigment. When at least one dominant allele
of this gene is present (**CC**, **Cc**) the agouti and black phenotypes can be expressed.
When this second gene is in the homozygous recessive form (**cc**) no pigment is
formed and the mice are albino.

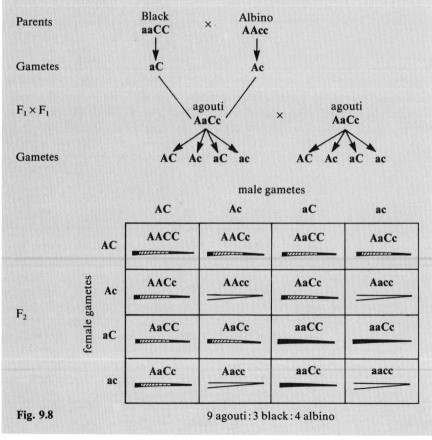

Fig. 9.8   9 agouti : 3 black : 4 albino

The gene for pigment development is therefore epistatic, when homozygous recessive, over both agouti (**AA, Aa**) and black (**aa**).

In a cross of pure-breeding black (**aaCC**) × pure-breeding albino (**AAcc**) mice the F_1 are all agouti. Interbreeding the F_1 produces a modified F_2 ratio of 9:3:4, agouti:black:albino. The 9/16ths that are agouti have at least one dominant allele of both genes. The 3/16ths black are homozygous recessive for **a** (**aa**) and have at least one dominant allele of gene **C**. The 4/16ths albino are the result of the epistatic action of the recessive homozygote, **cc** (Fig. 9.8).

Epistasis can also be seen in the inheritance of plumage colour in domestic fowl. Pure-breeding White Leghorns are homozygous dominant for a gene which causes inhibition of colour in the feathers (**II**). They are also homozygous dominant for an independent gene which brings about coloured plumage (**CC**). They are white because **I** is epistatic over **C**. White Wyandotte poultry carry the recessive alleles of the inhibition gene (**ii**) but are white because they are also recessive for coloured feathers (**cc**). When White Leghorns are mated with White Wyandottes (**IICC** × **iicc**) the F_1 are white (**IiCc**). Interbreeding among the F_1 gives an F_2 comprising white and coloured birds in 13:3 ratio. This comes about because all genotypes containing at least one dominant allele of the inhibition gene (**II, Ii**) are white and the double recessive (**iicc**) is also white due to lack of pigment. Only the **iiCC, iiCc** genotypes can be distinguished due to the presence of colour with no inhibition. If the Punnett Square is drawn out showing the F_2 genotypes the coloured birds will be seen to make up 3/16ths of the progeny.

There are numerous other cases cited in the literature of gene interactions and the various modified F_2 ratios that result from them. A comprehensive list is given in *Genetics* by M. W. Strickberger. The few examples we have given here will suffice to make the point that while the transmission of a pair of unlinked genes follows a simple and predictable pattern, their mode of action does not always do so. The idea that one gene controls one character is an oversimplified view of genetics which we take in the first instance for convenience and ease of explanation of basic principles.

Another complexity to which we must briefly refer here involves a single gene in the control of several different characters—a phenomenon known as 'pleiotropy'.

Pleiotropy

Pleiotropy is *the multiple phenotypic effect of a single gene.* Essentially there are two ways in which a single gene may affect several different characters. The first is that the gene produces a product which is involved in a branched biochemical pathway, and a mutation in the gene therefore affects different branches. In the pathway shown in Fig. 9.9, for example, enzyme α is necessary for step B–C. Mutation of the gene which determines α will result in a block in the pathway, and deficiency of end products E, F, G and H.

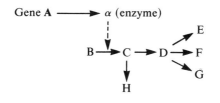

Fig. 9.9

The second cause of pleiotropy is when the gene determines an enzyme which is common to a number of different metabolic pathways. Mutation of the gene will then cause a block in these otherwise unrelated pathways (Fig. 9.10).

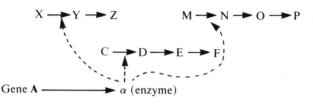

Fig. 9.10

An example is the 'mottled' mutation in the mouse. The primary defect is in the transport of copper across the intestinal lining, and the mice are copper-deficient. All enzymes requiring copper as a cofactor are defective and this leads to multiple effects including lack of pigment, curly whiskers, bone defects, arterial aneurisms (thin walls leading sometimes to internal bleeding) and neurological disorders.

Cytoplasmic inheritance

There are many exceptions to the rule in genetics. One of them is that not all inherited characters are determined by genes located in the nucleus. A small minority are controlled by genes located in cell organelles in the cytoplasm, i.e. **cytoplasmic genes**, and these of course are exceptions to the chromosome theory of inheritance. Since they are **extrachromosomal** (i.e. outside the chromosomes), such genes are not subject to the normal rules of Mendelian heredity.

Leaf variegation in plants

One of the earliest and best known examples of cytoplasmic inheritance is that discovered by Correns in a variegated variety of the four-o'clock plant *Mirabilis jalapa.* Variegated plants have some branches which carry normal green leaves, some branches with variegated leaves (mosaic of green and white patches) and some branches which have all white leaves (Fig. 9.11).

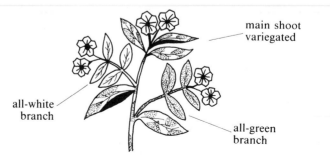

Fig. 9.11   Leaf variegation in *Mirabilis jalapa.*

Correns discovered that seed produced by flowers carried on the green branches gave progeny which were all normal green. It made no difference whether the phenotype of the branch which carried the flower used for pollen was green, white or variegated. Seed taken from white branches likewise gave all white progeny, regardless of the pollen donor phenotype. These of course died in the seedling stage. Seeds from flowers on variegated branches gave three kinds of progeny, green, white and variegated, in varying proportions; again regardless of the pollen donor phenotype. In other words, the phenotype of the progeny always resembled the female parent and the male made no contribution at all to the character. The effect is seen quite clearly in the difference which Correns found between reciprocal crosses:

$$♀ \text{ green} \times \text{white } ♂ \ \rightarrow \ \text{green progeny}$$
$$♀ \text{ white} \times \text{green } ♂ \ \rightarrow \ \text{white progeny}$$

The explanation for this unusual pattern of inheritance is that the genes concerned are located in the chloroplasts within the cytoplasm, not in the nucleus, and are therefore transmitted only through the female parent. In eukaryote organisms the zygote normally receives the bulk of its cytoplasm from the egg cell and the male gamete contributes little more than a nucleus. Any genes contained in the cell organelles of the cytoplasm will therefore show **maternal inheritance**. The leaf variegation is due to two kinds of chloroplasts: normal green ones and defective ones lacking in chlorophyll pigment. Chloroplasts are genetically autonomous (i.e. self-determining) and have their own system of heredity in the form of chloroplast 'chromosomes'. These are small circular naked DNA molecules which carry genes controlling *some* aspects of chloroplast structure and function. A mutation in one of these genes, which affects the synthesis of chlorophyll as in *Mirabilis*, will therefore follow the chloroplast in its transmission and will not be inherited in the same way as a nuclear gene—Fig. 9.12 and Box 9.2.

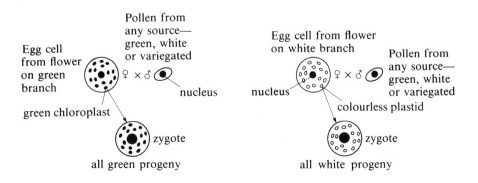

Fig. 9.12   Inheritance of leaf variegation in *Mirabilis jalapa*. The character is controlled by cytoplasmic genes located in the chloroplast 'chromosome'. Chloroplasts are self-perpetuating cell organelles and during sexual reproduction they are only transmitted through the cytoplasm of the egg cell, as undifferentiated proplastids. They are not inherited through the pollen. The progeny of crosses therefore have the characters of the female parent and show maternal inheritance.

The other important point to note about the inheritance of chloroplasts is that they have no regular means of distribution, such as chromosomes do at mitosis, whereby they can be equally shared out to the daughter cells following division. A plant that begins life as a zygote containing a mixture of normal and mutant chloroplasts cannot therefore maintain the same mixture in all of its somatic cells. The two kinds of plastids are shared out randomly during cell division, according to the way they happen to be placed in the cytoplasm when it is partitioned. Some branches of variegated plants may therefore remain mosaic while others, by chance, may turn out to contain all white or all green chloroplasts in all of their cells. In a similar way the flowers on a variegated branch may be of three kinds. Some will have egg cells with all green chloroplasts, some egg cells with all white and others will retain a mixture.

Box 9.2   The chloroplast genome

It is now known that the chloroplasts of plants carry their genetic information in the form of small circular DNA molecules, similar in size and form to the chromosomes of bacteria (Chapter 12). These DNA molecules contain genes which code for some of the proteins and RNAs used in chloroplast structure and function; and it is mutations in these genes which are most likely to be responsible for the leaf variegation effects described above. It must also be emphasised that chloroplasts are not totally independent of the nucleus in their heredity: most of their proteins are coded by nuclear genes, and mutations in these show normal Mendelian patterns of inheritance.

The DNA molecules which make up the chloroplast genome are 'naked' ones (Fig. 9.13) and bear no resemblance to the chromosomes of the nucleus, which are much larger and are composed of both protein and DNA (Chapter 2). The really surprising thing about the chloroplast DNAs is the large number of copies which are present: up to 300 in a mature plastid. Since an average of 160 chloroplasts are present in a mesophyll cell of the mature leaf of a cereal such as wheat, this means that there may be as many as 48 000 chloroplast 'chromosomes' per mesophyll cell. The reason for this enormous 'redundancy' of genetic information is unknown.

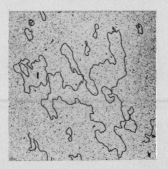

Fig. 9.13   Electron micrograph of a single circular molecule of chloroplast DNA (ctDNA) from the lettuce plant. The chloroplast 'chromosome' has a length of 155 000 base pairs of DNA. The small circles are 'chromosomes' of the virus ϕX174 (5370 nucleotides of DNA) which are included as a standard for calculating the size of the chloroplast 'chromosome'. (Photograph kindly supplied by Dr Tristan Dyer, Plant Breeding Institute, Cambridge.)

Other examples of cytoplasmic inheritance

Leaf variegation due to chloroplast mutation is known in numerous other genera of plants: *Epilobium* and *Pelargonium* are two examples.

Many of the other examples of cytoplasmic inheritance, in a variety of species, appear to involve characters which are associated with functions of the mitochondria. They have to do with defects in growth and ATP energy metabolism. Well known cases include the 'Poky' (slow growing) mutants in the fungus *Neurospora*, and 'Petite' mutants in brewers yeast. The mitochondria, like the chloroplasts, are self-replicating organelles which contain their own genes and have a limited number of characters which are independent of the nucleus. They are transmitted mainly through the female line and mutations in their genes show the same pattern of maternal inheritance. Mitochondrial 'chromosomes' have a similar circular configuration of 'naked' DNA as chloroplasts. In a typical haploid yeast cell each of the mitochondria contains in the region of 50 small circular 'chromosomes'.

Summary

Alleles of some genes interact with one another to give F_1 phenotypes that show incomplete dominance or co-dominance. Their heterozygotes can be distinguished from the homozygotes and this produces a modified F_2 phenotypic ratio of $1:2:1$. Deviations from the $3:1$ ratio also occur when one of the homozygous classes in an F_2 is determined by a lethal gene. When two different genes control the same single character they may also interact in their expression and give rise to modifications of the familiar $9:3:3:1$ ratio. There are various modifications, depending upon the kind of interaction that occurs, but the transmission of the genes is normal and the nine genotypes expected with independent segregation are all present. Genes which are located outside of the nucleus, in cell organelles, behave differently in reciprocal crosses and show a pattern of maternal inheritance.

Further reading

Jinks, J. L. (1976), *Cytoplasmic Inheritance*, Oxford Biology Reader No. 72, Oxford University Press.

Questions and problems

1 A parakeet having the allele **B** at one locus and **yy** at another locus develops blue plumage while an individual with **bb** at the first and **Y** at the second locus develops yellow plumage. An individual with both **B** and **Y** alleles develops green feathers, and one with the genotype **bbyy** is phenotypically white.
   (*a*) What are the meanings of the terms:
      (i) allele
      (ii) locus
      (iii) genotype?

(b) What is the significance of the convention concerning the use of capital and small letters for two alleles at a locus?

(c) Suggest why it has not been necessary to mention the second allele at one of the loci of each of the first two birds, above, and at both loci of the third bird.

(d) What is the simplest explanation of the green colour of parakeets that have both **B** and **Y** alleles?

(e) Below each of the four plumage colours make a list of *all* the possible genotypes.
(i) blue (ii) yellow (iii) green (iv) white.   (*O.L.E. Zool., 1979*)

2 A dominant allele **L** governs short hair in guinea-pigs and its recessive allele **l** governs long hair. Co-dominant alleles at an independently-assorting locus specify hair colour such that **YY** = yellow, **YW** = cream and **WW** = white. From matings between dihybrid short cream pigs **LlYW**, predict the phenotypic ratio expected in the progeny.

3 (a) What is meant by 'Mendelian inheritance'?

(b) Using one example of each, explain why (i) linkage, (ii) sex linkage and (iii) incomplete dominance do not give ratios characteristic of Mendelian inheritance.
(*L. Biol., 1979*)

4 The following data (from Sinnott, Dunn and Dobzhansky) show the results of a large number of matings carried out with a yellow variety of house mouse:

Parents	Offspring	
	yellow	non-yellow
Yellow × non-yellow	2378	2398
Yellow × yellow	2396	1235

How would you explain these data?

5 A recessive mutant allele of a gene responsible for the synthesis of chlorophyll in the tomato gives an albino plant when homozygous. The albino dies in the seedling stage after using up the food reserves in its seed. Heterozygotes give pale-coloured plants, but they survive. A normal green tomato plant was crossed with a heterozygote and the seed harvested. These seeds were planted and gave rise to progeny which were then self-pollinated. A further generation was grown using this selfed seed and the progeny were found to be a mixture of normal green and pale-coloured plants in a ratio of 5:2. Explain these results.

6 In poultry the genes for rose comb (**R**) and pea comb (**P**), if present together, produce walnut comb. The recessive alleles of both, when present together in the homozygous condition, produce single comb. What are the possible results that might be obtained by crossing a rose-combed fowl with pea comb? Explain your reasoning in full. What general principle is demonstrated here?

7 Radishes can be long, **LL**, round **ll** or oval **Ll**.

(a) If oval and long radishes grew close and could cross freely what would be (i) the shapes of the offspring, (ii) the ratio of shapes and (iii) their genetic constitution?

(b) How might round radishes be produced from oval?

(c) Explain why round radishes cannot be produced from long. If long radishes are crossed to round radishes and the F_1 are then allowed to cross at random among themselves, what phenotypic ratio is expected in the F_2?

8 Seed from a single tobacco plant was sown and produced a mixture of 149 green and 51 white seedlings. The white seedlings lacking chlorophyll soon died. Suggest a likely genetical explanation for these observations and explain what you could do to verify your hypothesis.

9 The autosomal gene for short tail (T) in the mouse is dominant to that for normal long tail (+). Two short-tailed mice of the same genotype were mated and the progeny in five successive litters were:

Litter number	1	2	3	4	5
Short-tailed mice	4	3	3	4	2
Long-tailed mice	2	2	2	1	2

From the progeny, eight short-tailed mice were mated with eight of their long-tailed siblings and produced a total of 28 short-tailed and 30 long-tailed mice.

(a) Explain the ratios obtained by working out the genotypes of the parents and the offspring in the two crosses.

(b) Comment on the effect of the short-tail gene (T) in mice.

(O.L.E. Biol., 1983, part question)

10
Multiple alleles and blood groups

In all of Mendel's experiments with peas the individual characters that he studied were each represented by two alternative forms. The single genes controlling these characters existed as two alleles, one of which was fully dominant over the other. The other examples of monohybrid inheritance that we have so far encountered have likewise involved genes with only two alleles, although we now know that the dominance between them may be modified by various forms of interaction (Chapter 9). There is no particular reason, of course, why a gene should have only two alleles. Genes are complex structures made up of linear sequences of hundreds of nucleotide pairs in DNA (Chapter 14) and they may mutate to give a number of different allelic forms, i.e. 'multiple alleles'. In this chapter we will deal with multiple alleles and also with the blood group genetics of man. Blood groups provide the best known examples of multiple alleles and it is therefore convenient to discuss the two topics together.

What are multiple alleles?

In diploids each gene is represented twice, once on each of the two homologous chromosomes at corresponding loci. A gene with two alleles may therefore have three different genotypes—two kinds of homozygote and a heterozygote (Fig. 10.1).

Fig. 10.1

With complete dominance there will be only two phenotypes, but in cases of incomplete dominance, and co-dominance, the heterozygote may also be distinguishable (Chapter 9).

The term **multiple alleles** refers to *the alleles of a gene which has more than two*. Suppose a gene has three alleles **W**, **w^a** and **w^b**, specifying three different forms of the same character. The gene is still represented only twice, at corresponding sites on the chromosome pair concerned, but we will now have more than three genotypes when we combine the alleles two at a time—in fact there are six possibilities (Fig. 10.2).

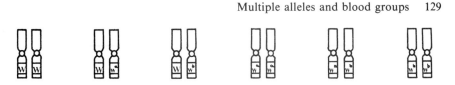

Fig. 10.2

The number of phenotypes will depend upon which alleles are dominant over which others and whether or not there is any incomplete dominance, or any co-dominance, involved as well. With four or more alleles the situation is correspondingly more complex. In experimental matings, however, we will generally be concerned with only two parents, and only two pairs of alleles, at any one time—as in a simple monohybrid cross.

Multiple alleles in man

ABO blood groups

The ABO blood group system is one of several different blood group systems in man that are genetically determined. Phenotypes are due to proteins which act as antigens bound to the surface of the red blood cells. In the ABO system the blood groups are controlled by a single gene, designated as I, which has three alleles, I^A, I^B and I^O. Allele I^A produces antigen A, I^B produces antigen B and I^O does not produce any antigen. I^A and I^B are co-dominant and both are dominant over I^O. There are six possible pair-combinations of the alleles, giving six genotypes within the human population, and four different phenotypes:

Genotypes	Antigens	Blood group phenotypes
$I^A I^A$	A	A
$I^A I^O$	A	
$I^B I^B$	B	B
$I^B I^O$	B	
$I^A I^B$	AB	AB
$I^O I^O$	—	O

Persons of blood group A carry antigen A and may be either homozygous for allele I^A ($I^A I^A$), or heterozygous ($I^A I^O$). Those of group B phenotype may likewise be homozygous ($I^B I^B$) or heterozygous $I^B I^O$ and they carry the B antigen. People in group AB are known to be heterozygotes ($I^A I^B$) and to carry both the A and B antigens, while those in group O have neither A nor B antigens on their red cells and are also of known genotype ($I^O I^O$).

It is a simple matter to say what phenotypes and genotypes are expected in the children of parents with various blood groups:

♀Group O × group O♂

$$\text{Parents} \quad ♀I^O I^O × I^O I^O♂$$

All of the children will be of blood group O.

♀Group AB × group O♂

A single child has a probability of $\frac{1}{2}$ of being group A and $\frac{1}{2}$ of being group B (Fig. 10.3).

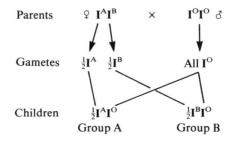

Parents	♀ $I^A I^B$	×	$I^O I^O$ ♂
Gametes	$\frac{1}{2}I^A$ $\frac{1}{2}I^B$		All I^O
Children	$\frac{1}{2}I^A I^O$		$\frac{1}{2}I^B I^O$
	Group A		Group B

Fig. 10.3

♀Group AB × group AB♂

For a single child the probability of being group A $=\frac{1}{4}$, group AB $=\frac{1}{2}$ and group B $=\frac{1}{4}$ (Fig. 10.4). You could work out for yourself the outcome of the various other kinds of crosses that can be made: A×A; B×B; A×B; A×AB; A×O; B×AB; B×O; and AB×O.

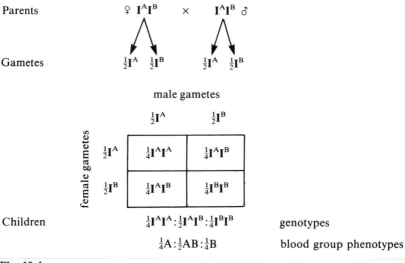

Parents   ♀ $I^A I^B$ × $I^A I^B$ ♂

Gametes $\frac{1}{2}I^A$ $\frac{1}{2}I^B$ $\frac{1}{2}I^A$ $\frac{1}{2}I^B$

male gametes

female gametes	$\frac{1}{2}I^A$	$\frac{1}{2}I^B$
$\frac{1}{2}I^A$	$\frac{1}{4}I^A I^A$	$\frac{1}{4}I^A I^B$
$\frac{1}{2}I^B$	$\frac{1}{4}I^A I^B$	$\frac{1}{4}I^B I^B$

Children $\frac{1}{4}I^A I^A : \frac{1}{2}I^A I^B : \frac{1}{4}I^B I^B$ genotypes

$\frac{1}{4}A : \frac{1}{2}AB : \frac{1}{4}B$ blood group phenotypes

Fig. 10.4

An understanding of the genetics of the ABO blood group series in man is of considerable practical importance in relation to blood typing and blood transfusion. These aspects are dealt with in Box 10.1.

Box 10.1   Blood typing and blood transfusion

Karl Landsteiner (1901) experimented with mixing blood from different people. He separated the red cells out by centrifugation and then mixed the cells of one person's blood with the serum of another. In some mixtures the serum caused the red cells to clump together, or **agglutinate**, and in others it did not. It was discovered that the clumping was caused by natural antibodies present in the serum which reacted with the A and B antigens on the surface of the red cells. Since a person cannot carry antibodies which would agglutinate his or her own cells, there are no anti-A or anti-B antibodies in the serum of people of group AB. Group O has no A or B antigens and carries both anti-A and anti-B antibodies; and groups A and B have antibodies against each other's antigens (Fig. 10.5). The agglutination reaction is shown in Fig. 10.6.

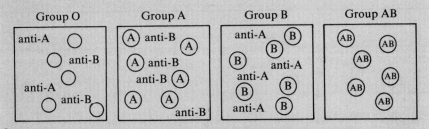

Fig. 10.5 Antigens and antibodies found in the four blood group types of the ABO series. ○ = antigen type on red cells; anti-A and anti-B = antibodies in serum.

Why do some people carry natural antibodies against one another's red cell antigens? The answer, it seems, is that these antigenic substances are not confined to red blood cells of human beings: they are common to many organisms, including the bacteria that are constantly in contact with our bodies. We are therefore exposed to them in our day-to-day lives and make antibodies against them in the same way as we would with any other foreign proteins that entered our bodies. Antigens of the same type as a person has on his/her own red cells are not recognised as foreign, and no antibodies are made against them.

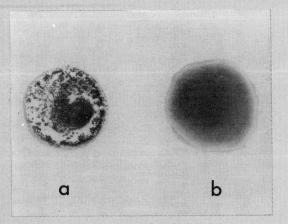

Fig. 10.6 (a) Agglutination of the red cells will occur when a drop of blood from a person of group A is mixed with serum from a person of group B. The B serum contains anti-A antibodies which react with the A antigens on the surface of the red cells and cause them to clump together.

(b) When the same blood is mixed with the serum of another person of type A there is no agglutination.

Blood typing

The existence of natural antibodies makes blood group classification (typing) a simple matter. Two kinds of serum are used for making tests. Serum containing anti-A antibodies is prepared from a person of blood group B, by removing the cells, and serum with anti-B antibodies is prepared from the blood of an individual of group A. A small drop of the blood to be typed is mixed with the two kinds of serum and the clumping reaction noted. A person whose blood is not aggluti-nated by either kind of serum is group O (i.e. has no antigens). One whose cells are agglutinated by anti-A, but not anti-B, is group A (has *A* antigens). Conversely cells agglutinated by anti-B, but not anti-A, belong to group B (has *B* antigens). An individual whose cells are clumped by both kinds of serum is in group AB (has *A* and *B* antigens). The reactions are summarised in Fig. 10.7.

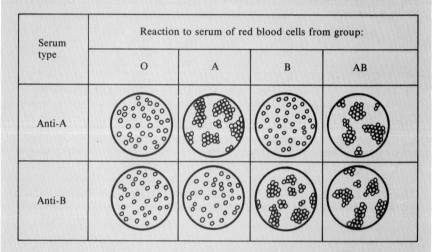

Serum type	Reaction to serum of red blood cells from group:			
	O	A	B	AB
Anti-A				
Anti-B				

Fig. 10.7   Blood groups are typed according to the way their red cells react when mixed with two kinds of serum containing anti-A and anti-B antibodies.

Blood transfusion

Blood can be *safely* given by transfusion from one person to another only when both donor and recipient are of the same group. If the groups do not match, there is a danger that the red cells of one person, or both of them, will be clumped by the antibodies of the other's serum. The reaction that matters most is that which occurs between the donor's red cells and the recipient's serum. The reason is that the donor usually contributes a small quantity of blood which is mixed with a much larger volume of that of the recipient. If the recipient's serum contains antibodies against the donor red cells, then because of the unequal quantities the incoming red cells will be completely agglutinated. They will cause blockage of the capillaries and may result in the death of the recipient. If the donor's serum has antibodies against the recipient's red cells, the effect is much less damaging. This is because the incoming antibodies are diluted out in the large volume of the recipient's blood. They are also absorbed by various other tissues of the body and quickly removed from the circulation.

When small quantities of blood are transfused some groups can therefore be mixed with others, as explained in Table 10.1.

Table 10.1 Summary of agglutination reactions of a donor's cells which occur when blood of different groups is mixed during transfusions. Agglutination is denoted by a plus sign and these combinations have to be avoided. The donor's antibodies are not shown because the important factor in transfusion is the way the donor's red cell antigens react with the recipient's antibodies—as explained in the text.

	Recipient			
Donor	O anti-A anti-B	A anti-B	B anti-A	AB
O	–	–	–	–
A (*A*)	+	–	+	–
B (*B*)	+	+	–	–
AB (*AB*)	+	+	+	–

Group O can donate to any of the others because O has no *A* or *B* antigens on its red cells and cannot be agglutinated by antibodies in the recipient's serum. For this reason people of group O are referred to as **universal donors**.

AB blood has no anti-A or anti-B antibodies and can receive blood from any of the other groups: it is unable to agglutinate their red cells. Persons of group AB are known as **universal recipients**.

Blood of group A cannot be given to people of groups O or B, since the anti-A antibodies in their serum would clump the incoming cells immediately. Similarly B type blood is not donated to groups O or A.

MN blood groups

The MN blood groups are another case of multiple alleles in man. They are determined by a different gene from those of the ABO system. In introductory texts it is usual to confine the discussion of them to the two main alleles and to make only passing reference to the fact that there are many subdivisions of alleles within the main ones—as indeed there are in the ABO system. As mentioned in Chapter 9, the two alleles are co-dominant. Each one produces an antigen on the red cells so that homozygotes (**MM, NN**) have either the *M* or *N* antigen and the heterozygotes (**MN**) have both. There are thus three different genotypes in the population and three blood group phenotypes which correspond to them:

Genotype	Antigen	Phenotype
MM	*M*	M
NN	*N*	N
MN	*MN*	MN

Inheritance of the gene is straightforward. The six different crosses that can be made are given below together with their outcome in the progeny:

Parents	Children
MM × MM	All MM
NN × NN	All NN
MM × NN	All MN
MM × MN	$\frac{1}{2}$ MM : $\frac{1}{2}$ MN
NN × MN	$\frac{1}{2}$ NN : $\frac{1}{2}$ MN
MN × MN	$\frac{1}{4}$ MM : $\frac{1}{2}$ MN : $\frac{1}{4}$ NN

There are no natural antibodies in serum against the *M* and *N* antigens—and therefore no problems with transfusions. To classify MN blood types it is necessary to prepare two kinds of antiserum (anti-M and anti-N) by injecting human red cells into rabbits to induce antibody formation. Rabbit blood containing the antibodies is then used to obtain purified antiserum which can be used in agglutination tests.

Rhesus blood groups

Rhesus blood groups were described by Landsteiner and Wiener in 1940 when they found that antiserum prepared by injecting blood of the rhesus monkey into rabbits and guinea pigs could agglutinate the red cells of man. Eighty-five per cent of people tested were shown to carry an antigen which reacted with the rhesus antibodies, and 15% did not. Those with the antigen are called 'rhesus-positive' types and those without are 'rhesus-negative'. The difference was later shown to be due to a single gene with two alleles, symbolised as Rh^+ and Rh^-. Rh^+ is fully dominant over Rh^- so that heterozygotes have the rhesus-positive phenotype:

Genotypes	Antigen	Phenotypes
Rh^+Rh^+	+	rhesus-positive
Rh^+Rh^-	+	rhesus-positive
Rh^-Rh^-	−	rhesus-negative

Inheritance of the gene follows the typical Mendelian pattern for a single gene with two alleles. Rhesus-negative parents have only rhesus-negative children. The outcome of matings between two rhesus positives depends upon whether the parents are both homozygous, both heterozygous or one of each:

Parents	Children
$Rh^+Rh^+ \times Rh^+Rh^+$	All Rh^+Rh^+
$Rh^+Rh^+ \times Rh^+Rh^-$	$\frac{1}{2}$ Rh^+Rh^+ : $\frac{1}{2}$ Rh^+Rh^-
$Rh^+Rh^- \times Rh^+Rh^-$	$\frac{1}{4}$ Rh^+Rh^+ : $\frac{1}{2}$ Rh^+Rh^- : $\frac{1}{4}$ Rh^-Rh^-

It is now known that the alleles of the **Rh** gene form a multiple allelic series and that suitable immunological tests can distinguish many subdivisions of the two main alleles—but we will not concern ourselves with these details.

One of the most important aspects of the Rh antigens is their involvement in the blood disorder **erythroblastosis fetalis** which occurs when the rhesus blood types of a mother and her foetus are incompatible. The details are explained in Box 10.2.

Box 10.2   Erythroblastosis fetalis

The condition of erythroblastosis fetalis arises when a rhesus-negative mother (Rh^-Rh^-) carries a foetus which has an Rh^+ allele transmitted by the father. The foetus is then heterozygous (Rh^+Rh^-) and produces rhesus antigens on its red blood cells; i.e. it has the rhesus-positive phenotype. Leakage of blood across the placenta, if some damage occurs at or near to birth, can allow rhesus-positive cells from the foetus' blood into the mother's circulation—and she then builds up antibodies against them in her blood serum. The mother thus becomes 'sensitised', and when she carries a second rhesus-positive child her antibodies may be present in high enough concentration to agglutinate (clump) and break down the red cells of the foetus. Such antibodies can pass freely across the placenta from mother to foetus. The problem is not as great as might be thought, however, since the father is rhesus-positive and the mother rhesus-negative in only about one in eight marriages; and then only one in 200 infants are affected by the disease. It is also now possible to give protection against the condition by injecting the mother with anti-Rh serum immediately after the birth of the first rhesus-positive child. This treatment destroys any rhesus-positive cells that may have entered her blood while she was carrying the child, and prevents the build up of antibodies in her system.

For reasons which we cannot go into here, this kind of incompatibility between mother and foetus is far less troublesome for the other blood group systems.

Legal aspects of blood group genetics

Blood groups are useful characters for resolving legal cases about such problems as baby 'mix-ups' in maternity hospitals, disputed paternity cases and forensic science. The main reason is that the characters are so distinctive, easily classified and determined by single genes with simple patterns of inheritance. Their importance in relation to disease and blood transfusion also means that extensive records exist of the blood types of a large number of the population, whereas this kind of information would not necessarily be available for other phenotypic characters. Another factor which is of importance in forensic science is that the blood group of a person can be determined from traces of blood left behind at the scene of a crime long after the perpetrator has gone. In the case of the ABO blood groups it is also possible to classify the blood type without using an individual's blood at all. Some people are 'secretors' and produce the *A* and *B* antigens in certain other body fluids such as saliva, tears and semen.

The main point about using blood groups as evidence is that they can never prove that an individual was involved in a particular misdemeanour, but they **can** definitely *exclude* him or her. If a young lady claimed, for instance, that a certain gentleman was the father of her child, and she was blood group O, the baby was O and the accused AB, then clearly he would be innocent. If his blood group was O, on the other hand, this would not necessarily prove that he was the father—because there are numerous other males in the population with that blood group.

Example A farmer has two sons. The first, born when the farmer was young, grew into a handsome healthy youth, in whom he took great pride. The second, born much later, was always a sickly child and neighbours' talk induced the farmer to bring his wife to court disputing its paternity. The grounds of the dispute were that the farmer, having produced so fit a first son, could not be the father of a weakling.

The blood groups were:

	ABO	MN
Farmer	O	M
Mother	AB	N
First son	A	N
Second son	B	MN

What advice can we give the court?

If we look at the ABO system, the farmer was O and therefore $I^O I^O$ in genotype. The mother was AB and therefore $I^A I^B$ in genotype. The cross was therefore:

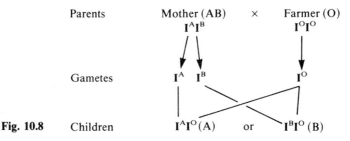

Fig. 10.8   Children

We can see that they can have sons who have A or B blood types, and on this basis both of their sons could have come from the marriage.

When we consider the MN system, however, we see that the farmer must have been **MM** (as he was phenotype M) and the mother must have been **NN** (as she was phenotype N) so the children can only be **MN**:

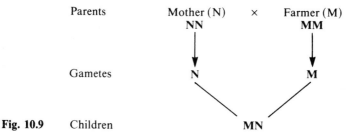

Fig. 10.9   Children

Since the first son was N blood type (and therefore NN genotype) we would have to tell the farmer that he could have fathered the second son but that he was *definitely not the father of the first.*

In addition to the three genes discussed above there are at least twelve others in man which determine lesser known, and less important, blood group systems, such as **Kell, Lutheran, Duffy, Kidd, Lewis** etc. They are all controlled by different independently-inherited genes with two or more alleles. These blood group genes have their own specific antigen types, all of which are bound to the surface of the same red blood cells. When considered all together, they make it possible to describe a really detailed 'genetic profile' of the blood group phenotype of an individual person.

Other examples of multiple alleles

In practical genetics we often confine ourselves to studying only two forms of a character, controlled by two alleles, because this is sufficient to determine the pattern of inheritance of a gene and to find out how it is linked to other genes on the chromosome map. When any individual gene is studied in detail it usually turns out to have more than just two alleles. Multiple alleles seem to be the rule rather than the exception.

The white eye locus in *Drosophila* has at least twelve different alleles specifying twelve different eye colours in between the normal red and the white. An individual fly will carry only two of the alleles at the locus concerned, as homozygous or as various heterozygous combinations.

Another well known example of multiple alleles is in the gene which determines coat colour in rabbits. The gene is designated as C and it exists in four different allelic forms: C = 'agouti'; c^{ch} = 'chinchilla'; c^h = 'himalayan'; c = 'albino'. Phenotypes of the pure breeds are as follows:

1. **Agouti (CC):** This is the full colour of the wild rabbit. Agouti is an overall greyish-brown colour produced by the grey hairs which have a band of yellow on the shaft, as in agouti mice (Fig. 10.10*a*).
2. **Chinchilla ($c^{ch}c^{ch}$):** The coat resembles that of the real chinchilla (i.e. the South American rodent). It is grey in colour with a 'ticking' of black hairs. The yellow band of the agouti is replaced with a band of pearl (Fig. 10.10*c*).
3. **Himalayan** (c^hc^h): The himalayan has a white coat except for the extremities—nose, ears, feet and tail—which are black. The eyes have pigment, unlike those of the albino (Fig. 10*d*).

   Development of pigment in the fur of the extremities is due to a temperature-sensitive enzyme which works only below a certain critical temperature. Dark fur can be developed in other areas of the body if these areas are exposed to cold treatment, such as ice-pack, during growth of the fur. Baby himalayans have no dark pigment—it only develops after they leave the nest. A similar gene causes the familiar colouring in Siamese cats. This particular mutation provides an excellent example of the way in which the genotype and environment interact to produce the phenotype.
4. **Albino (cc):** Albinos have no pigmentation at all. Their fur is pure white and the eyes pink (Fig. 10.10*b*).

The dominance relationships between the four alleles are

$$C > c^{ch} > c^h > c$$

(a)   (b)

(c)   (d)

Fig. 10.10   Coat colour phenotypes in the rabbit. (a) Agouti, (b) albino, (c) chinchilla, (d) himalayan. ((a)-(c) photographed at Whipsnade Zoo, by courtesy of the curator; (d) courtesy of Mr J. C. Sandford.)

Crossing between the breeds in all possible pairwise combinations gives ten different genotypes, but only four phenotypes because of the dominance effects:

Phenotypes	Genotypes
Agouti	CC; Cc^{ch}; Cc^{h}; Cc
Chinchilla	$c^{ch}c^{ch}$; $c^{ch}c^{h}$; $c^{ch}c$
Himalayan	$c^{h}c^{h}$; $c^{h}c$
Albino	cc

F_1 monohybrids therefore display the phenotype of the allele which is the more dominant of the two, and the F_2 will segregate out in the familiar $3:1$ Mendelian ratio (Fig. 10.11).

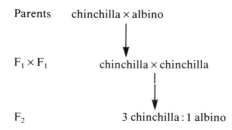

Parents   chinchilla × albino

$F_1 × F_1$ chinchilla × chinchilla

F_2 3 chinchilla : 1 albino

Fig. 10.11

Summary

Genes may have more than two alleles. As a result of mutation they can have several different forms, i.e. multiple alleles, which determine several different phenotypes for a single character. The best known example is the gene controlling the ABO blood group series in man, which has three main alleles. Another well known case is the gene for coat colour in rabbits which has four alleles. When a gene is represented by multiple alleles, only two of them may be present at the locus concerned in any one individual.

Questions and problems

1 After eight years of married life, during which time she had failed to become pregnant, Mrs X met and fell in love with Mr Y. During the ensuing five years three children were born. In the meantime the persons involved had tried to come to an understanding and wished to determine which of the two men was the father of each child. The blood types of those involved were determined with the following results:

Person	ABO	MN
Mrs X	O	MN
Mr X	O	MN
Mr Y	A	N
First child	O	MN
Second child	O	M
Third child	A	N

What can we conclude as to the paternity of each child?

2 A mother in a maternity hospital claimed that the baby (a) allocated to her was not her own. There was only one other baby (b) in the ward at the time. The mother is blood group O and MN, and cannot taste phenylthiocarbamide (PTC). Baby (a) is blood group A, M and can taste PTC. Baby (b) is O, MN and cannot taste PTC. The husband of the mother is now dead, but she has three other children:

(i) A, MN, taster
(ii) B, N, taster
(iii) A, MN, non-taster.

Which baby do you think belongs to her? Give your reasoning.

3 The inheritance of coat colour in cattle involves a multiple allelic series with a dominant hierarchy as follows:

$$S > s^h > s^c > s.$$

The S allele puts a band of white colour around the middle of the animal and is referred to as a 'Dutch belt'; the s^h allele produces 'Hereford' type spotting; solid colour is a result of the s^c allele; and 'Holstein' type spotting is due to the s allele. Homozygous Dutch-belted males are crossed to Holstein-type spotted females. The F_1 females are crossed to Hereford-type spotted male of genotype $s^h s^c$. Predict the genotypic and phenotypic frequencies in the progeny.

4 (*a*) Most people possess erythrocytes which incorporate an antigen called the rhesus factor. Explain why, when a rhesus-negative woman has children fathered by a rhesus-positive man, the first child is usually normal but the second child and subsequent children may suffer from erythroblastosis fetalis (haemolytic disease of the new-born).

(*b*) Four babies were born in the same hospital. Due to a power cut, their identities were confused before they could be given to their parents. The babies' blood groups were A, B, O and AB. The blood groups of the parents were:

Mr and Mrs Farmer	A × B
Mr and Mrs Miller	B × O
Mr and Mrs Draper	O × O
Mr and Mrs Carter	AB × O

Name the parents of each of the babies and give the reasoning which enabled you to obtain the answers. *(A.E.B. Biol., 1983)*

11
Continuous variation

In introducing genetics, we have so far confined our studies to characters which are controlled by one or two genes and which have distinctive and clear-cut alternative forms. The alleles of the genes concerned have major effects upon the phenotype which are readily distinguished and which cannot be confused with any additional variation which may be due to the environment. This kind of clear-cut difference in the forms of various characters is known as **discontinuous variation** (e.g. tall and short peas). It is essential to begin the study of genetics in this way in order to identify individual genes and to explain the basic principles of the subject in the clearest possible terms. At the same time it is important to emphasise that in nature, and outside of the experimental laboratory, characters of the visible phenotype which show only a few discontinuous and easily classified forms are the exception rather than the rule. Most of the natural variation that we see about us, and which is thought to be the most important in terms of natural selection and evolution, and in agriculture, is not of this discontinuous kind at all. If we look at the more obvious variations within a species such as their size, weight and morphology, we see that these characters seldom fall into categories that are easily classified, or grouped, into any simple pattern. On the contrary they are represented by numerous forms, and though there may be a range of sizes, from small to large, the intermediates differ from each other by very little, and so form a range of what is known as **continuous variation**. Characters of this kind are much more difficult for the geneticist to work with. They are usually determined by several genes, each of which has a small effect, and which are not readily identified as individual units of heredity. In this chapter we will briefly explain how we can study the inheritance and the genetic basis of continuous variation, and see why it is that we normally give so little attention to it in our elementary studies. At the end of the chapter we will also summarise everything that we have so far encountered about the relation between genes and characters.

Continuous and discontinuous variation

It will be helpful at the outset to explain precisely what we mean by the terms 'continuous' and 'discontinuous' variation. We will deal with discontinuous variation first because this is the pattern of variation with which we are already familiar. If we examine a population of peas of mixed heights, say from the backcross Tt × tt, we will find the situation depicted in the histogram in Fig. 11.1a. The plants fall into two discrete and discontinuous classes without any intermediate forms. They are either tall or dwarf (Fig. 4.4). We have seen the same clear-cut pattern of variation in the pea seed-colour phenotypes, of yellow

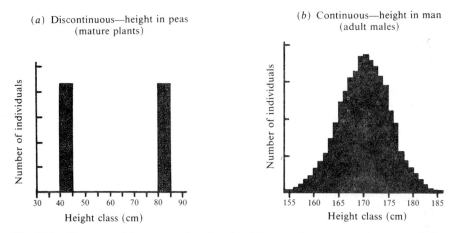

Fig. 11.1 Frequency histograms showing the difference between continuous and discontinuous patterns of variation.

(a) Height in peas is a character which shows discontinuous variation. Plants fall into two discrete and quite separate groups and individuals can be easily classified as either 'tall' or 'dwarf'.

(b) The character height in man shows a more continuous range of variation. There are numerous forms which are not so discrete, or separate, but which blend into one another to give a more continuous range of distribution. If the height classes are measured in intervals which are small enough, the pattern of distribution becomes virtually a smooth curve.

and green, and in the white, pink and red flower colours of the four-o'clock plant (Chapter 9). Many other similar examples can be listed—yellow and purple seeds in maize, horned and polled (hornless) cattle, vestigial and normal wings in *Drosophila*, brown and blue eye colour in man, and so on. Characters which fall into a few easily classified types in this way, without a range of intermediates, are also said to display **qualitative variation**. The different forms are recognised by their **qualities**, such as 'tallness' and 'dwarfness', and we don't have to make any measurements in order to classify them. There may be some minor variations between one tall pea plant and another, and also among the individual short ones, which are due to environment; but because the difference in the two character forms is determined by the alleles of a single **major gene**, with large effects, we have no difficulty in distinguishing between them.

Continuous or **quantitative variation**, on the other hand, means that the character has numerous forms which are not discrete, or separate, but which all merge into one another to give a continuous distribution of values. 'Height' in man is an example. In a population such as that shown in the frequency histogram in Fig. 11.1b there are no discrete types to be found with respect to this character. We cannot place individuals into only two or three categories of tall, short or intermediate height, because they do not fall into such a simple pattern of variation. There is a continuous distribution of heights spread throughout the range 150-190 cm. The bulk of the population is grouped around the mean height value of 170 cm with the rest tailing off on either side in a symmetrical pattern. If enough measurements are taken they can be divided into so many height classes that the histogram becomes a smooth line curve.

A symmetrical frequency histogram of this kind is known as a **normal curve**. The population can be described in terms of the **mean** and the **standard deviation** (i.e. the spread of the histogram). To describe any one individual in the population shown in Fig. 11.1*b* we would have to measure his height and classify him in terms of cm and mm. In other words the character is **quantitative** and individuals have to be **quantified** (measured, in this case) in order to describe their phenotype. The vast majority of characters of living organisms are of this quantitative kind and show a pattern of continuous variation which fits a normal curve. Another well known example is the character of 'intelligence' in man. Others include the weight and size of most plants and animals, the yield of crop plants, the egg-laying capacity of hens, milk yield in cows, etc. It is difficult to deal with the genetics of quantitative characters, simply because the variations do not fall into a few simple classes, as they do with qualitative variations, and we cannot identify the individual genes which control them (unless they are very few in number, p. 146). Generally speaking, characters which show continuous variation are controlled by a large number of different genes, each of which has only a small effect. Their individual contributions are therefore 'lost' against the background of environmental variation, and it becomes difficult in most cases to sort out the influence of environment from that of genotype.

Johannsen's pure line experiment

How do we know that quantitative characters are controlled by genes?

The Danish geneticist Johannsen gave us the answer to this question with his detailed breeding experiments on the dwarf bean, *Phaseolus vulgaris*.

The character which Johannsen studied was seed weight. He chose to work with the dwarf bean because it is a self-fertilising plant and all of the descendants of one seed are what he called a 'pure line'; i.e. all of the seeds taken off one plant are genetically identical. Pure lines are homozygous at all their gene loci and they breed true. Different lines are homozygous for different combinations of dominant and recessive alleles, e.g. **AAbbccDDee** ... or **aaBBCCddEE** ... A more detailed explanation is given in Box 11.1.

Johannsen began his experiments with a collection of 19 pure lines obtained from different sources. By using lines of diverse origin he knew that they would be genetically distinct. Each line had a characteristic mean seed weight. It ranged from 35.1 centigrams (cg) in line 19 to 64.2 cg in line 1. Within the lines there was variation in seed weight due to environmental factors, such as the position that a seed occupied within the pod and the position of the pod on the plant.

In one experiment Johannsen mixed together seeds from all 19 lines. A frequency histogram of their weights gave a continuous distribution over the range 5–95 cg, with an overall mean of 48 cg. Although the mean weights of the lines all differed from one another, the variation within lines, due to environmental effects, was so large that it 'masked' these mean differences and so gave rise to the continuous pattern of variation. He then demonstrated that it was possible to change the mean seed weight in progeny grown from samples of this mixture by *selection*. He picked out two samples, one of small seeds and the other of large seeds. These were grown and the progeny seeds they produced were weighed. The progeny differed in their mean weights. Those coming from the 'small parents' were lighter than those from the 'large parents' (Fig. 11.2*a*).

When the same selection procedure was repeated using the variation *within lines* there was no response (Fig. 11.2*b*). Johannsen concluded from this that variation in the character of seed weight could be separated into two components: **heritable** and **non-heritable variation**. Seeds from different lines vary as a result of genetic and environmental factors. Selection on the mixture works because in the process of picking out large and small seeds different genotypes are being taken, and it is the heritable part of the variation which is being transmitted to the progeny to give the difference in mean seed weight between the two samples. Seeds from the same pure line have no genetic differences. All their variation is due to environment and is non-heritable. Progeny from samples of large and small seeds within lines have identical mean weights because their parents have the same genotype.

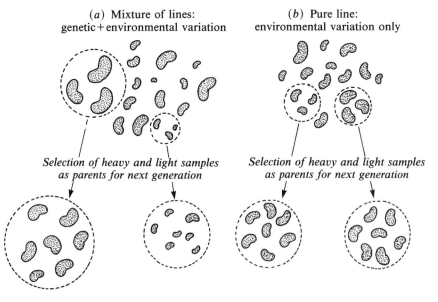

(*a*) Mixture of lines: genetic + environmental variation

(*b*) Pure line: environmental variation only

Selection of heavy and light samples as parents for next generation

Selection of heavy and light samples as parents for next generation

Progeny showing response to selection: mean weights significantly different

Progeny showing no response to selection: mean seed weights identical

Fig. 11.2   Johannsen's experiment with the dwarf bean showed that quantitative characters are controlled by genes and that with inbred pure lines it is possible to distinguish between heritable and non-heritable variation.

Box 11.1   Pure lines

Pure-breeding lines are found in plants which are self-fertilising. When an individual is homozygous for the alleles of a single gene (**AA**) it will produce only one kind of gamete (**A**) in both its ovules and pollen grains. By self-pollination and fertilisation the identical gametes will combine (**AA**) and the progeny will breed true. This same argument can be extended to cover as many genes as we like—**AAbbCCDDeeffGG** ... As long as cross pollination is prevented, by natural means or otherwise, the inbred line will remain pure. Lines which come from diverse sources may differ in the homozygous combinations of the pairs of identical alleles which they carry—e.g. **AAbbccDDEEFF** ..., **aabbCCddEEff** ... If a pure line becomes heterozygous at a locus, by mutation or outcrossing, then repeated self-fertilisation over a small number of generations will quickly restore the homozygosity:

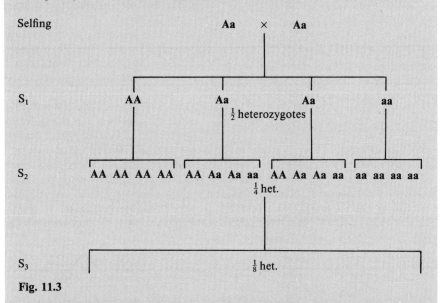

Fig. 11.3

The heterozygotes become an ever-diminishing fraction of the progeny and eventually they are 'bred out'. We finish up with two homozygous pure lines (**aa** and **AA**). An individual that is heterozygous at many loci, such as an outbreeding plant being forcibly 'selfed', will go through the same process but it will take longer and we will end up with a large number of different pure lines.

The importance of Johannsen's work was in showing that for a quantitative character part of the variation at least is controlled by genes, and that by using the appropriate experimental procedure it is possible to 'dissect' the variation into its two component parts and to distinguish between what he called the **genotype** and the **phenotype**. The genotypic component is that which is represented by the mean seed weights of the lines, and the phenotypic part is the variation about the mean within lines.

In outcrossing species (e.g. cross-pollinating species of plants) there are no pure lines and every individual is a different genotype (Chapters 7 and 18). In these cases it is obviously much more difficult to unravel the two components of continuous variation.

Johannsen showed quite convincingly that quantitative variation was due to genes as well as to environment. But he wasn't able to relate the differences between his lines to any *particular* genes, or to identify any individual units of heredity.

Multiple factors

Following Johannsen's work there was controversy among geneticists as to whether or not continuous variation could be accounted for on the basis of Mendel's idea of discrete units of inheritance. Some favoured the suggestion that the blending of phenotypes could easily be explained if the character was controlled by several different genes each with a small cumulative effect. Others argued that Mendel's unit factors were not the answer and that some other, entirely different, type of hereditary factor was involved.

The issue was resolved in about 1910 when firm evidence for the **multiple factor**, or **multiple gene**, hypothesis was provided by the Swedish geneticist Nilsson-Ehle, and subsequently by several others. Nilsson-Ehle worked with grain colour in wheat (*Triticum aestivum*). In one of his experiments he crossed a pure-breeding variety with dark red grains to one with colourless white grains (wheat is also a self-fertilising species). The F_1 had grains of an intermediate colour. In the F_2 there was a ratio of 15 coloured:1 white, but the coloured grains varied in the density of their red pigmentation. The character was not *qualitative* because the F_2 could not be described as simply red or white. It was *quantitative* and each grain had to be classified on a scale of five shades of colour from white through to dark red. As the shades were not easily distinguished it could be argued that the variation was practically continuous. By selfing the F_2 plants, and looking at their breeding behaviour in the F_3, Nilsson-Ehle was able to confirm their genotypes and to show that the character was controlled by only two genes (Fig. 11.4).

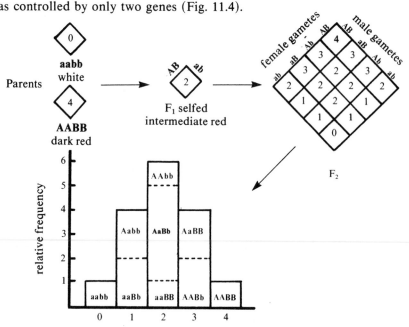

Fig. 11.4 Nilsson-Ehle's experiment on the genetic basis of grain colour in wheat. A cross between two pure-breeding strains which differ by two genes with additive (cumulative) effects gives an almost continuous range of variation in the F_2. Each dominant allele contributes one unit of colour density to the character.

The genes concerned behaved in their transmission like any other Mendelian genes. The only difference in this instance was that the two genes worked in such a way that their effects were **cumulative** or **additive**, and each dominant allele contributed a certain degree of redness to the colour expression of the phenotype. In the model shown in Fig. 11.4 we have assigned an arbitrary value of one unit to the colour contribution of each dominant allele and a value of zero to the recessives. On the basis of **additive gene action** it is now a simple matter to see how the independent segregation of only two pairs of alleles from a heterozygous F_1 can give a range of five phenotypic classes in the F_2. If an element of environmental variation is included as well it becomes even easier to appreciate how a character controlled in this way, by relatively few genes, can approximate to a continuous distribution.

It is important to understand that in this experiment we are dealing with two genes determining *one* character, not two separate characters as in Mendel's work, and we do not expect to find a $9:3:3:1$ ratio in the F_2. The situation is also different from that of duplicate gene action (in the shepherd's purse), which we discussed in Chapter 9, where the character had only two forms and the action of either, or both, dominant alleles gave the same phenotype. Here the genes are acting in an *additive* manner and we get an effect upon the phenotype which is directly proportional to the number of dominant alleles present in the genotype.

In practice, of course, most cases of continuous variation are not so amenable to study as this one, and in most cases we have no idea how many different genes are involved and no simple means of finding out. The value of Nilsson-Ehle's experiment, and others like it, lay in emphasising how continuous variation could be understood in terms of simple Mendelian genetics. There is no need to invoke anything more than multiple genes to explain it. The word **polygenes** is now used in preference to multiple genes to describe the genes which control a quantitative character. The term **polygenic inheritance** was introduced by Mather to describe *characters whose expression is controlled by many genes with individual slight effects upon the phenotype*. The main features of polygenic inheritance are summarised as follows:

1. The characters are controlled by a number of genes.
2. The genes have individual small effects and changing one allele for another one at a locus causes relatively little difference in the phenotype.
3. The phenotype is subject to considerable environmental variation.
4. Characters show a continuous range of variation.

Genes and characters

At this stage in introducing genetics it may be helpful to collect together our ideas on the relation between the gene and the character that we have so far encountered, since it is obviously not as simple as we suggested in some of the earlier chapters. In the summary given overleaf, it is assumed that the inheritance of the genes is according to Mendelian principles of segregation and independent segregation and that what we are concerned with here are the different forms of action and interaction of genes, at the level of the visible phenotype, rather than any complications in their transmission.

Genes and characters

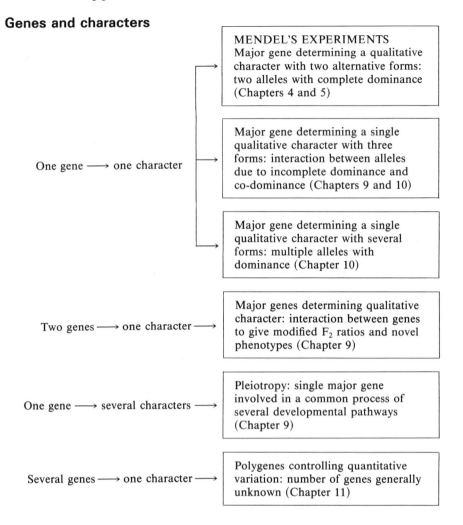

One gene ⟶ one character

MENDEL'S EXPERIMENTS
Major gene determining a qualitative character with two alternative forms: two alleles with complete dominance (Chapters 4 and 5)

Major gene determining a single qualitative character with three forms: interaction between alleles due to incomplete dominance and co-dominance (Chapters 9 and 10)

Major gene determining a single qualitative character with several forms: multiple alleles with dominance (Chapter 10)

Two genes ⟶ one character ⟶
Major genes determining qualitative character: interaction between genes to give modified F_2 ratios and novel phenotypes (Chapter 9)

One gene ⟶ several characters ⟶
Pleiotropy: single major gene involved in a common process of several developmental pathways (Chapter 9)

Several genes ⟶ one character ⟶
Polygenes controlling quantitative variation: number of genes generally unknown (Chapter 11)

The summary is not meant to be taken as a complete list of all possible modes of gene action, but is given only to put together those aspects that we have covered so far. We will have more to say about gene action in later chapters when we come to deal with the utilisation of genetic information at the molecular level (Chapters 14 and 15).

Summary

Characters which show continuous variation are controlled by several genes (polygenes). The individual contribution of the genes to the phenotype is small, relative to that of the environment, and it is difficult to identify them as separate units of inheritance. Where this can be done the experiments suggest that the number of polygenes is not necessarily high and that they are transmitted according to the same laws as major genes.

Questions

1 (a) By reference to a named species, illustrate what is meant by (i) continuous variation and (ii) discontinuous variation.

   (b) Explain how such variation may arise.   *(C. Biol., 1982)*

2 A cross was made between two maize (*Zea mays*) varieties called Tom Thumb and Black Mexican which differed markedly in ear length. The ear length of both parents, the F_1 and the F_2 generations, was measured to the nearest centimetre. This is given in the table below with the number of ears in each length category (for example, 14 of the F_1 plants produced ears 12 cm in length).

	Ear length (cm)																
	5	6	7	8	9	10	11	12	13	14	15	16	17	18	19	20	21
Black Mexican parents									3	11	12	15	26	15	10	7	2
Tom Thumb parents	4	21	24	8													
F_1					1	12	12	14	17	9	4						
F_2					1	10	19	26	47	73	68	68	39	25	15	9	1

   (a) Both parents showed variation in ear length. The variation in the F_1 is comparable to the average of the parental variations while the variation in the F_2 is greater than that found within the parental lines or the F_1. How would you explain this?

   (b) How would you explain the whole spectrum of different degrees of expression of this particular characteristic?

   (c) What is the term used to describe the range of the phenotypes in the above example?   *(C. Biol., 1983)*

12
Bacteria and viruses

Geneticists classify living organisms into two main groups, **eukaryotes** and **prokaryotes**, according to the way in which their cells are organised. Eukaryotes are plants and animals. They have a true nucleus (= eukaryote) enclosed within a membrane and cells which contain complex organelles. In addition they have a nucleolus, several pairs of chromosomes composed of a complex of DNA and protein, and processes of mitosis and meiosis.

The prokaryotes (= before the nucleus) include the bacteria and blue-green algae. Their cellular organisation is much simpler. They have only one 'chromosome', which consists of a single circular molecule of DNA which is found in a central position within the cell without any surrounding membrane. They have no mitosis or meiosis. The viruses are below the level of organisation of prokaryotes and are treated as a separate group. They are non-cellular and consist simply of a single molecule of nucleic acid (DNA or RNA) which is usually enclosed within a protein coat. The structure of DNA is discussed in detail in Chapter 13.

Up till now we have been concerned only with the genetics of eukaryotic organisms, and have made no reference at all to the very different genetic systems that are found amongst the prokaryotes and the viruses. The reason for this is that the basic principles of genetics were mainly worked out with higher plants and animals during the first part of this century, and the bacteria and viruses made little contribution to the subject before 1950. Before that time there was little understanding of how prokaryotes and viruses related to living organisms, and little knowledge either of how to use them experimentally. Nobody knew how to mate one bacterium (or virus) with another, or what characters to work with.

Once the technical difficulties of handling microbes were overcome, their value as experimental organisms was quickly appreciated and they came to play a major role in the development of the science of genetics. The advances which came about from their use include much of what we now know about the structure of the gene and about the action, regulation and mutation of the genetic material at the *molecular* level (Chapters 14–17). The genetic code was elucidated using bacteria and viruses, and these micro-organisms are now playing a central part in the rapidly expanding fields of research into gene cloning and genetic engineering.

In this chapter we will give a brief account of the genetics of bacteria and viruses. The intention is to provide some idea of the way in which their heredity and variation is organised and to see how it differs from that of the eukaryotes. This knowledge will be helpful as a foundation to some of the experiments which are described in several of the forthcoming chapters, and also as a basis for the account of genetic engineering which is given in Chapter 22.

Bacteria

Background

Bacteria are very small unicellular organisms which can be seen as individuals only under the light, or electron, microscope. They are abundant in nature and have a great diversity of forms and different kinds of metabolism. Their cell structure is relatively simple in comparison with that found in higher plants and animals. They possess neither the rigid cellulose cell wall which is typical of plants, nor any chloroplasts or mitochrondria, but they do have ribosomes. The general structure of a bacterial cell is shown diagrammatically in Fig. 12.1.

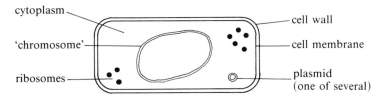

Fig. 12.1   Diagram showing the general structure of a bacterial cell. The 'chromosome' consists of one circular molecule of DNA, and is represented here in a highly simplified form. In reality the 'chromosome' is very long, in relation to the size of the cell, and is folded into a compact body called a nucleoid. The DNA in the 'chromosome' is 'naked', i.e. it is not associated with protein molecules, and does not form the type of chromosome which is present in eukaryotes (Chapters 2 and 13). Although the bacterial 'chromosome' consists of only one molecule of DNA it is often referred to, and represented, as *double-stranded*. This is because a DNA molecule is made up of two strands (i.e. two chains of nucleotides), as explained in Chapter 13. Plasmids are small circular DNA molecules which may be present in the cell in addition to the main bacterial 'chromosome'. Each plasmid also consists of one 'naked' double-stranded DNA molecule.

Bacteria are excellent organisms with which to study genetics. They are unicellular, relatively simple in structure and biochemistry and they are without many of the complexities of development and differentiation that are associated with multicellular species. Their life cycle is incredibly short—as little as forty minutes in some cases. They can be cultured and handled within the laboratory and several millions of them can be grown up overnight in a single bottle of culture medium. They can be grown in liquid medium and then spread (or 'plated') out on solid agar, on which they are immobilised, in such a way that each of the colonies which form represents a clone of several thousand genetically identical individuals derived by binary fission from a single cell; Fig. 12.2 overleaf.

Characters of bacteria   What characters can we use to study the genetics of bacteria?

The form of the colony is one character. In some species there are strains in which the individual cells are **encapsulated**, i.e. enclosed, in a polysaccharide capsule and this gives rise to colonies which are smooth in outline and distinguishable by eye from the 'rough' strain which lacks the polysaccharide (p. 159). The character of the colony is representative of the individual bacterium from which it arose by cell division. Another visible phenotype is the presence or absence of a flagellum: this can be assessed by the form that the colonies take when they are grown on soft agar. When the bacteria are mobile the colony appears as a 'flare', i.e. it grows as a streak across the agar rather than as a discrete spot.

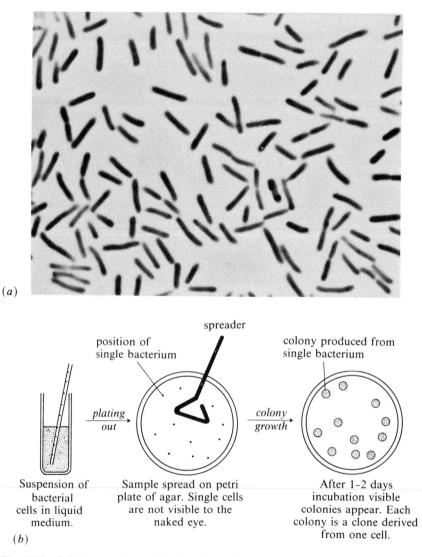

(a)

(b)

| Suspension of bacterial cells in liquid medium. | Sample spread on petri plate of agar. Single cells are not visible to the naked eye. | After 1–2 days incubation visible colonies appear. Each colony is a clone derived from one cell. |

Fig. 12.2 (a) Suspension of *Escherichia coli* bacteria growing in liquid medium. Photographed under the light microscope (×1000 magnification).

(b) Bacteria are easily cultured in the laboratory by growing them in a liquid nutrient medium. When a sample of the cell suspension is diluted and spread on a solid nutrient agar each of the cells will multiply by binary fission to form a small colony. Each of the colonies is a clone and has the genotype and phenotype of the single cell from which it originated.

The most widely used characters with which the bacterial geneticist works are those involving **biochemical mutations**. These are strains having single enzyme deficiencies which result in the loss of certain biochemical functions, e.g. the inability to synthesise a particular substance. As we will see in Chapter 14 most genes work by determining the structure of a particular enzyme, so that the basis of most character differences is likely to involve defects in enzyme activity.

The nutrition of normal wild strains of bacteria is referred to as **prototrophic**: they can grow on a **minimal medium** which contains only water, a suitable carbon source (often glucose) and a mixture of inorganic salts. From these simple ingredients they manufacture all the substances that they require for cell growth and multiplication. Gene mutations (Chapter 15) can cause 'blocks' in biochemical pathways resulting in mutant bacteria that are deficient in some aspect of their biochemistry. Such biochemical mutants are unable to synthesise certain substances, such as particular amino acids, nucleotides or vitamins, and these therefore have to be supplied to the culture medium in order for the cells to grow. The mutants are said to have a 'growth requirement', because they are unable to grow on a minimal medium, and are called **auxotrophic mutants**. They have to be cultured on a complete medium (made from an extract of yeast and hydrolysed protein) which contains a mixture of all possible substances required for growth, or on a minimal medium supplemented with the particular deficient substance (i.e. amino acid or vitamin) concerned. It is a simple matter for the bacterial geneticist to isolate large numbers of different biochemical mutants with which to study the structure and action of the genes, and the way in which they are organised within the bacterial 'chromosome'. The procedure is essentially the same as that which was first used to obtain biochemical mutants in the fungus *Neurospora*, which is explained in Chapter 14. Another class of mutants which are widely used are the **resistant mutants**, which display resistance to certain antibiotics. Normal wild strains are sensitive and will die if antibiotics are included in the culture medium. This class of mutant is also easily obtained. All that is needed is to plate out several million cells of a wild-type sensitive strain onto a medium containing an antibiotic such as penicillin. Mutations giving resistance to penicillin occur spontaneously in about one in 10^6 cells and these mutant cells will then survive and give rise to colonies. The resistant mutant strain can then be isolated from the colony and established as a pure strain.

Chromosome organisation and replication

Bacteria have all of their genes organised into one linkage group. They are haploid and their single 'chromosome' consists of one circular double-stranded molecule of 'naked' DNA (Fig. 12.1). It does not form a complex with any histone protein, as in eukaryotes (p. 181), and neither is it found within a membrane-bound nucleus. It is found as a dense body called the **nucleoid** within the cytoplasm. In the bacterium *Escherichia coli*, which inhabits the digestive tract of man, the 'chromosome' has a total length of more than a millimetre, and it is folded into a nucleoid which is less than one micron long. The 'packing' ratio is therefore of the order of 1000 : 1. Surprisingly, this long double-stranded molecule of DNA is able to unravel itself and to make a copy by self-replication (p. 179) in as little as 40 minutes. The way in which the 'chromosome' replicates and is distributed into the daughter cells is illustrated in Fig. 12.3. In addition to the main bacterial 'chromosome' there may be one, or more, small circles of double-stranded DNA known as **plasmids**, which are present in the cytoplasm of the cell (Fig. 12.1). Their significance will become apparent later. When plasmids are present they replicate themselves independently of the main chromosome, and do so in a similar manner.

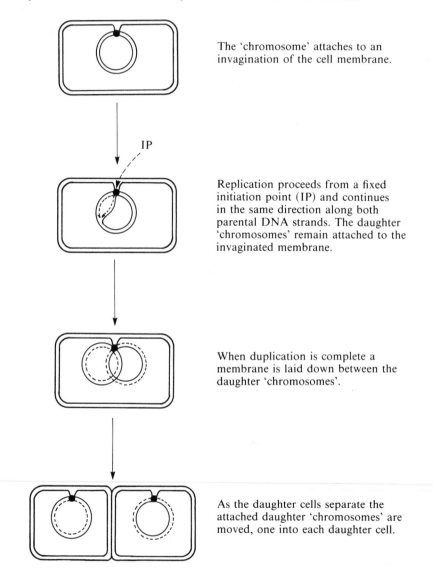

The 'chromosome' attaches to an invagination of the cell membrane.

IP

Replication proceeds from a fixed initiation point (IP) and continues in the same direction along both parental DNA strands. The daughter 'chromosomes' remain attached to the invaginated membrane.

When duplication is complete a membrane is laid down between the daughter 'chromosomes'.

As the daughter cells separate the attached daughter 'chromosomes' are moved, one into each daughter cell.

Fig. 12.3 Diagram showing a simple model of how the replication and distribution of the bacterial 'chromosome' takes place during cell division. The circle represents a molecule of double-stranded DNA which in reality is more than a thousand times the length of the cell. Details of the structure and replication of DNA are given in Chapter 13.

Recombination in bacteria

At one time it was thought that mutation (Chapter 15) was the only means by which genetic variation arose in bacteria. They have no obvious visual differentiation into male and female forms and no obvious ways in which they can exchange their genes with one another. Painstaking research with auxotrophic mutants, however, revealed that bacteria can have fascinating kinds of sex life and several different ways of undergoing genetic recombination. The processes that we will describe serve as a natural means of generating variation. They also provide a method for genetic studies, because the occurrence of these processes means that crossing experiments can be carried out.

Lederberg and Tatum (1946) were the first to show recombination between two strains of bacteria. They worked with what is now the most widely used and important of all the species in genetics, *Escherichia coli*. They used two auxotrophic strains, both of which were unable to grow on minimal medium. Strain Y10 carried mutations in three genes so that it was unable to make the amino acids threonine (thr^-) and leucine (leu^-) or the vitamin thiamine (thi^-). Strain Y24 had three different mutations, and could not synthesise the amino acids phenylalanine (phe^-) and cysteine (cys^-) or the vitamin biotin (bio^-). Each strain carried the normal forms of the genes (denoted by a '+' superscript) for which the other one was mutant.

$$\text{Strain Y10} \quad thr^- leu^- thi^- phe^+ cys^+ bio^+$$
$$\text{Strain Y24} \quad thr^+ leu^+ thi^+ phe^- cys^- bio^-$$

These two strains (i.e. the parental types), neither of which was able to grow by itself on minimal medium, were mixed together for a short while in a liquid culture of complete medium and then a sample of the culture was plated out on a minimal medium. Surprisingly, some colonies grew and these turned out to be prototrophic recombinants combining the normal forms of all six genes—three from one parent and three from the other:

mixture of auxotrophic parental strains in complete medium

$$\text{Y10} \quad thr^- leu^- thi^- phe^+ cys^+ bio^+ \; \times \; thr^+ leu^+ thi^+ phe^- cys^- bio^- \quad \text{Y24}$$
$$\downarrow$$
$$thr^+ leu^+ thi^+ phe^+ cys^+ bio^+$$

(recovery of some prototrophic recombinants able to grow on minimal medium)

After eliminating all the various forms of experimental error, such as contamination, transformation (see p. 159) and mutation, which could have accounted for the result, Lederberg and Tatum concluded that the two strains must have 'mated' together in some way and exchanged their genes by recombination—but they didn't know how. Later experiments confirmed their findings and showed that physical contact between the two strains was essential for recombination to occur. A large number of elegant experiments led to an understanding of the process involved, which became known as **conjugation**.

Conjugation   It is now known that in *E. coli* there are two mating types which determine a primitive form of sexuality. There are **donor cells**, which donate genetic material, and **recipients** which receive it during 'mating'. The difference between the two is due to the presence of a plasmid called the **sex factor**, or **fertility factor**, which is given the symbol **F**. It is a small circular DNA plasmid about 1/40th the size of the main 'chromosome'. Cells lacking the F-factor are designated as F^- and are recipients. Those with the F-factor are the donors and they may exist in one of two alternative states.

In one of these states, known as F^+, the F-factor is present in the cytoplasm in a 'free' form, i.e. it is inherited as a normal plasmid, independently of the main 'chromosome'.

In the other state, known as **Hfr**, it is integrated into the main 'chromosome' and transmitted along with it during replication and cell division (Fig. 12.4).

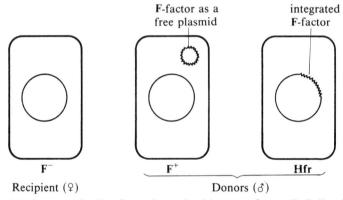

F-factor as a free plasmid   integrated F-factor

F⁻   F⁺   Hfr

Recipient (♀)   Donors (♂)

Fig. 12.4   Mating types in *E. coli* are determined by a sex factor, F. Cells without the F-factor are classified as F⁻ and are the equivalent of females in other species: these receive genetic material from F⁺ and **Hfr** types which are the donors and the equivalent of males in other species. (The bacterial 'chromosome' and the F-factor are shown as single strands for simplicity.)

The change in state from F^+ to **Hfr** is a reversible one brought about by a single crossover event following close pairing of the F-factor with a homologous region of the main 'chromosome'. For simplicity the bacterial 'chromosome' and the F-factor plasmid are represented here by a single line (Fig. 12.5).

bacterial 'chromosome'   F-factor 'free'   F-factor integrated →**Hfr strain**

Fig. 12.5

There are several different locations in the main bacterial 'chromosome' at which the F-factor may become inserted, to give several different **Hfr** strains. The F^+ and **Hfr** donor cells develop hair-like appendages. When donor and recipient cells find themselves in close proximity in culture the appendages of the donor make physical contact with the recipient and a conjugation tube is formed which connects the cytoplasm of the two mating cells. The outcome of this contact is different for $F^+ \times F^-$ and $Hfr \times F^-$. In $F^+ \times F^-$ matings only the F-factor is transferred between donor and recipient. $Hfr \times F^-$ matings, on the other hand, involve transfer of the main chromosome and it is this type of cross which leads to recombination and which is most useful to the geneticist. **Conjugation** is defined as *the unidirectional transfer of genetic information between donor and recipient cells involving direct cellular contact.*

F^+ and F^- cells   When F^+ and F^- cells conjugate, the F-factor in the donor cell replicates, and a copy of it is donated to the F^- recipient which then becomes an F^+ type as well (Fig. 12.6).

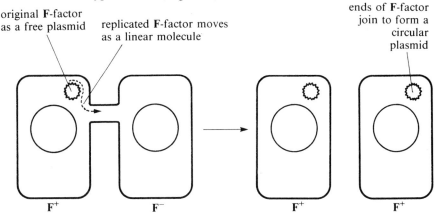

Fig. 12.6

In this way the F-factor can rapidly spread throughout an entire population of F^- cells, changing them all into F^+. The reason why F^- cells persist in a population is that the F-factors are sometimes spontaneously lost during division of F^+ cells, possibly because they fail to replicate themselves as fast as the bacterial 'chromosome'.

Hfr and F^- cells   **Hfr** stands for high frequency of recombination. In a mixture of **Hfr** and F^- strains the two kinds of cell conjugate in pairs and the **Hfr** partner donates a copy of its 'chromosome' to the F^- recipient. The process is described in Fig. 12.7. Transfer is in one direction only and the 'chromosome' passes through the conjugation tube as a linear molecule after opening up at a special site within the inserted F-factor. Part of the inserted F-factor then leads the way into the recipient cell. Only rarely does the entire 'chromosome' pass through the conjugation tube, and when this happens the remaining part of the F-factor is the last piece of the 'chromosome' to be transferred. In most matings the conjugation tube breaks during transfer. The sex of the recipient therefore remains as F^- and recombination can only involve that part of the donor 'chromosome' that has gained entry into the recipient's cell (Fig. 12.7).

In bacteria we cannot see the details of the crossing over process in the DNA, in the way that we can see chiasmata in the chromosomes of eukaryotes. The entire cell of *E. coli* (2 μm) is about a quarter of the size of one metaphase chromosome in an onion.

The events depicted in Fig. 12.7 are what we deduce is happening from our genetic studies based on the types and frequencies of the various recombinants that we find. Genotypes of the recombinants in the mating mixture are identified by growing samples of the culture on several differently supplemented media to see which substances they can and cannot manufacture. The recombinant

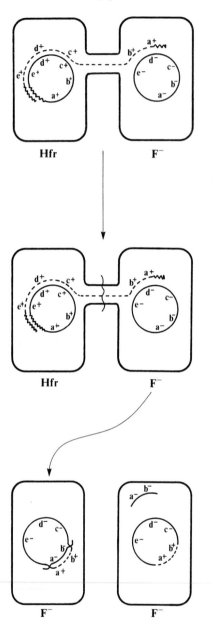

A conjugation tube is formed between the **Hfr** and the **F⁻** cell. The **Hfr** 'chromosome' makes a copy of itself which is then transferred into the **F⁻** cell. Transfer begins with the smaller end of the F-factor.

'Chromosome' transfer is often incomplete due to breakage of the conjugation tube. In those mating pairs in which the whole 'chromosome' does manage to pass through the tube the larger main part of the F-factor is the last part to be transferred.

Homologous segments of the donor and recipient 'chromosomes' pair together and recombination occurs. It involves the incorporation of genes from the donor cell into the recipient's chromosome. The small fragment released from the recipient is lost (it doesn't replicate) during the subsequent cell divisions.

Fig. 12.7 Diagram showing the way in which recombination takes place following conjugation between **Hfr** and **F⁻** cells of the bacterium *E. coli*. For simplicity the 'chromosome' is shown as a single strand. The parental cells at the start of mating are of genotype $a^+b^+c^+d^+e^+$ (and **Hfr**) and $a^-b^-c^-d^-e^-$ (and **F⁻**). The recombinant recipient cell has the genotype $a^+b^+c^-d^-e^-$ and has different biochemical characters from either of the two parents going into the 'cross'. Other kinds of recombinants can arise in mating pairs in which different lengths of the **Hfr** donor 'chromosome' are transferred before the conjugation tube is broken.

$a^+b^+c^-d^-e^-$ would be unable to make substances denoted by the symbols c, d and e, and could not grow on a minimal medium which lacked them; but it would be able to synthesise a and b and grow in their absence.

Because the 'chromosome' is transferred in a linear manner, starting at one end, it is possible to map the genes. This is done by an **interrupted mating** experiment in which a sample of the culture is removed at intervals of time after the start of mating, and the conjugating pairs separated by violent agitation. The time sequence in which the various 'marker' genes appear as recombinants gives their order along the 'chromosome'.

Two other means of genetic exchange between bacteria are known: besides conjugation, there are 'transformation' and 'transduction'. Unlike conjugation, these two systems do not require physical contact between the participating cells.

Transformation is *the exchange of genetic information brought about by the uptake of naked DNA by a recipient cell.* The mechanism by which it takes place is now well understood, and it is known that quite large fragments of DNA can easily pass through the wall of the bacterial cell and undergo recombination with homologous segments of the recipient's 'chromosome'.

The principle of transformation was discovered by Griffith in 1928, in the bacterium *Diplococcus pneumoniae*, which causes pneumonia in mice and various other mammals including man. At that time the nature of the transforming substance, i.e. the DNA, was not known. Griffith worked with two strains of *D. pneumoniae*. One of them was a virulent form, designated 'S', which could be recognised in culture by the smooth appearance of its colonies. This phenotype is caused by a polysaccharide substance which encapsulates the cells and protects them from the host (mouse) defence mechanism. It caused pneumonia when injected into mice. The other strain, 'R', was avirulent (i.e. non-virulent) and did not cause pneumonia. It lacks the polysaccharide capsule and gives colonies which have a 'rough' appearance when grown on agar plates. The two strains could thus be distinguished by the visible morphology of the colonies when grown in culture as well as by their effect upon the host organism:

S = smooth, virulent (causes pneumonia)
R = rough, avirulent (does not cause pneumonia)

Griffith used these two strains of bacteria in an unusual experiment. He injected the living R strain into a number of mice and found that it had no ill effect. Heat-killed R were likewise avirulent. When live S cells were injected the mice succumbed to pneumonia and died. Heat-killed S bacteria had no ill effect. The interesting idea was the injection of mice with both dead (heat-killed) S and live R. Surprisingly, they suffered high mortality, and when an autopsy was performed, living cells of both the S and R strain were recovered from their blood. The experiment is illustrated in Fig. 12.8. How was it that *live* S cells could be recovered from mice which had been injected with *dead* S and live R cells?

Griffith concluded that a substance, which he called the **transforming principle**, had passed from the remains of the dead S cells into the R strain cells and altered their heredity. The R had somehow been **transformed** into S by the presence of the dead S cell fragments. It wasn't until 1944 that Avery,

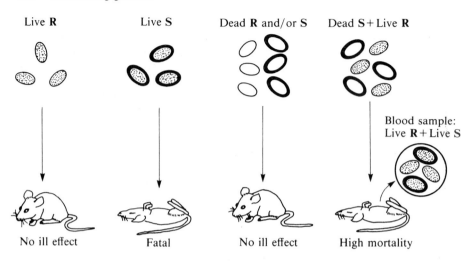

Fig. 12.8 Griffiths' transformation experiment with *Diplococcus pneumoniae*. Living bacteria are shown with cell contents and heat-killed dead ones are represented as empty cells. The smooth S-strain can be distinguished from the rough **R**-strain by the presence of its polysaccharide 'coat'.

MacCleod and McCarty identified the transforming substance as pieces of DNA. They isolated and purified the DNA from S-type *D. pneumoniae* and showed that it alone, and no other cell constituent, was effective in causing transformation. Their experiment was among the first to demonstrate that DNA is the genetic material, as we will discuss in Chapter 13.

The process of transformation is thought to come about by recombination following the entry of DNA fragments from the dead S-type remains into the living R cells. Some of the DNA fragments carry the gene, or genes, which determine the virulence of the S strain and they become incorporated into the chromosome of some of the avirulent R cells changing them into S (Fig. 12.9).

We will have more to say about transformation in Chapter 22. The phenomenon is not confined to bacteria; it can happen with plant and animal cells as well.

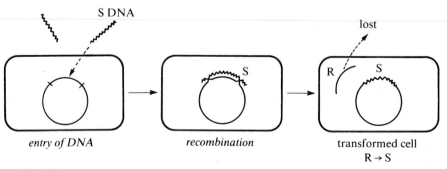

Fig. 12.9

Transduction is *the transfer of genetic material from one bacterium to another by a virus.* It was discovered by Lederberg and Zinder in 1951. They observed that recombination could take place between two auxotrophic strains of the bacterium *Salmonella typhimurium* even when the two strains were separated in the two sides of a U-tube by a bacterial filter. They found that the culture medium in the tube contained the virus P22 which was small enough to pass through the filter, and they concluded that the virus was carrying bacterial genes from one strain to the other. Transformation was ruled out by including enzymes in the medium to destroy any naked DNA. Transduction is explained more fully in the next section after we have described the structure and reproduction of viruses.

Viruses

Viruses are *sub-microscopic particles composed of a molecule of DNA (RNA in some cases) and usually encapsulated in a protein coat.* They are entirely dependent on the cells of other living organisms for their reproduction. Their life cycle consists of an intracellular phase, in which they reproduce, and an inert extracellular phase in which their nucleic acid is protected by the protein coat. In their extracellular state viruses show an immense variety of form and structure—some examples are given in Fig. 12.10. Although viruses use plants and animal cells, as well as bacteria, as hosts we will confine our remarks here to the bacterial viruses only, because of their importance in genetics. *A virus which depends upon infection of a bacterial cell for its reproduction* is known as a **bacteriophage** (or **phage** for short).

Structure and reproduction of bacteriophages

The general structure of a bacteriophage particle is shown in the diagram in Fig. 12.11. The single linear molecule of double-stranded DNA which is enclosed in the head portion is about 500 times longer than the head itself. As in the bacteria, it is difficult to envisage how the packaging of the DNA is achieved.

Bacterial viruses reproduce by injecting their DNA into a bacterial cell and then using the biochemical facilities of the host cell to make more copies of themselves. The tail fibres enable the virus to attach to the bacterial cell wall, which is then punctured. The injection of DNA is achieved by contraction of the protein sheath. What happens once the viral 'chromosome' has entered the host cell depends upon whether the phage is virulent or temperate.

Virulent phages always reproduce immediately, resulting in **lysis** (bursting) of the cell and release of progeny viruses. The reproductive cycle of a virulent bacteriophage is shown in a simplified diagram in Fig. 12.12.

Temperate phages do not always kill their host cells. They may enter a lytic cycle of development and behave as the virulent phages do, or else they may integrate their DNA into the main bacterial 'chromosome' as a **prophage**. In the latter event they replicate passively as part of the host 'chromosome' and are transmitted along with it into the descendent cells. A bacterium which carries a prophage is immune to any further infection and is said to be **lysogenic**, i.e. it has the heritable property of being able to lyse should physiological conditions change and the prophage be induced to come out of the host chromosome. Induction may be brought about experimentally, for example by irradiation with UV light. The life cycle of a temperate phage is summarised in Fig. 12.13.

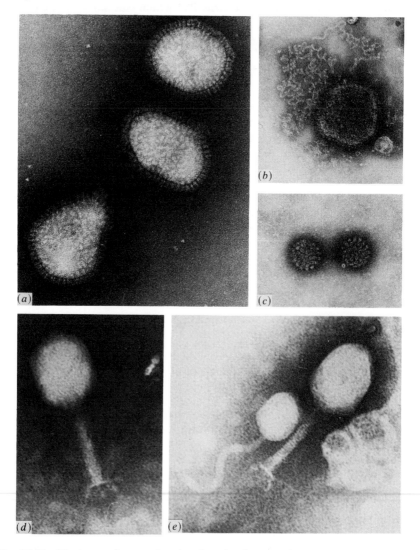

Fig. 12.10 Electron micrographs showi.ig the form and structure of several virus particles. (*a*) Influenza virus ×260 000. (*b*) Rabies virus ×168 000: the particle has been disrupted spilling out its 'chromosome'. (*c*) Human wart virus ×195 000. (*d*) T4 bacteriophage of *E. coli* ×250 000 (approx.). (*e*) T4 (right), and B3 bacteriophage of the bacterium *Pseudomonas*, ×250 000 (approx). (Photos (*a*)-(*c*) by courtesy of Dr Elwyn Griffiths, National Institute for Biological Standards and Control, Hampstead, London. Photos (*d*) and (*e*) by courtesy of Professor Don Ritchie, Genetics Dept., University of Liverpool.)

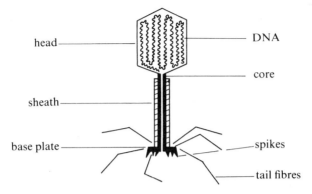

head ——————————— DNA

—— core

sheath ———————

base plate ——————— spikes

—————— tail fibres

Fig. 12.11   Structure of bacteriophage T2. The head contains a single molecule of DNA and is attached to a core surrounded by a contractile sheath. At the basal end of the core is a hexagonal base plate containing 6 short spikes and 6 long tail fibres.

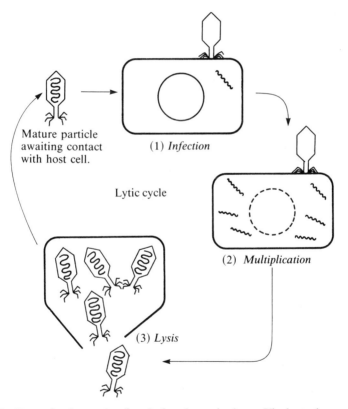

Mature particle awaiting contact with host cell.

(1) *Infection*

Lytic cycle

(2) *Multiplication*

(3) *Lysis*

Fig. 12.12   Reproductive cycle of a virulent bacteriophage. Virulent phages undergo reproduction immediately upon entering the host cell (1). After 5-10 minutes the bacterial chromosome begins to break up, and the viral DNA starts replicating and making more copies of itself (2). After 20-30 minutes new protein coats have been manufactured and assembled around the phage DNA (3). The cell is then lysed and the mature phage particles released.

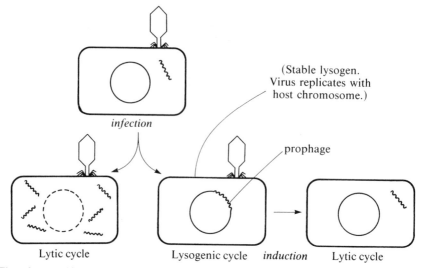

Fig. 12.13   Life cycle of a temperate bacteriophage.

Transduction

As we have already mentioned, viruses can **transduce** bacteria, i.e. transfer genes from one bacterium to another. In this process it is the genes of the bacteria which are recombined, not those of the virus. There are two different ways in which transduction can occur, corresponding to the two ways in which bacteriophages reproduce themselves, i.e. 'generalised' and 'specialised' transduction.

Generalised transduction   When virulent phages reproduce within a host bacterium, their new protein coats sometimes get wrapped around a small piece of bacterial 'chromosome' instead of around their own DNA. This happens with a frequency of about one virus particle in every 100 000 that are produced—so it is a rare event. Following lysis the 'aberrant' phage can then infect another bacterium, because the property of infection is determined by the protein coat and not by the viral DNA. In this way a piece of bacterial DNA (as much as 20–30 genes) can be transferred from one bacterium to another with the phage as a vehicle. Following infection the transferred fragment may pair with the homologous region of the main 'chromosome' and undergo recombination. As the viral coat may pick up any part of the bacterial 'chromosome', and transport it into another cell, this process is known as **generalised transduction**. The principal events are summarised in Fig. 12.14.

Specialised transduction by temperate phages   In the life cycle of temperate phages there is a lysogenic phase in which the virus resides as a prophage at a specific place within the main bacterial chromosome (p. 161). Following induction the prophage is released and may enter a lytic cycle. Some of the released phages are *defective*: they come out of the bacterial 'chromosome' in a slightly different form to that in which they were originally integrated. They carry with them a small piece of the bacterial 'chromosome' and leave some of their own DNA behind. These defective phages may then infect another host cell and in so doing they will transfer the attached bacterial genes to another bacterium. As the prophage is always integrated into the same position within the host 'chromosome' there are only a few bacterial genes that can be carried with it as a 'hitch-hiker'. For this reason this kind of transduction is said to be **specialised**, or **restricted** (Fig. 12.15).

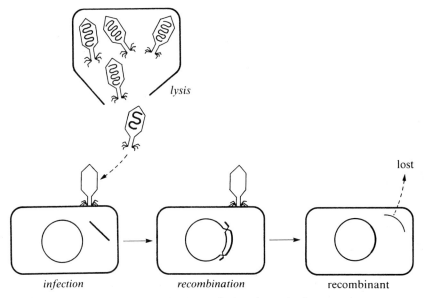

Fig. 12.14   Generalised transduction by a virulent bacteriophage.

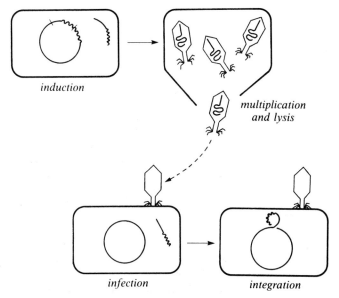

Fig. 12.15   Specialised transduction by a temperate bacteriophage.

Genetics of bacteriophages

Bacteriophages evidently play an important part in the genetics of bacteria. The way in which they carry genes between one bacterial cell and another is a most useful aid to experimentation, apart from any significance that it may have in nature. But what of the phages themselves? How do we study their genetics? They are after all too small to be seen as individual particles, either with the naked eye or with the light microscope. What phenotypes, or characters, can we work with and how do we cross one virus with another?

One of the main characters used by geneticists is that of **plaque morphology**. When a dilute suspension of phages is spread out onto a 'lawn' of bacteria growing on an agar plate, individual virulent phages will infect any bacterial cells that they happen to be in contact with and will lyse them. The progeny phages that are released will infect neighbouring cells and will lyse them as well. As a result a clear area, or **plaque**, will appear in the lawn of bacteria (in a period of 12 hours) in the region where the bacterial cells have been destroyed. Plaques may vary in their morphology (small, large, fuzzy or sharp at the edges, etc.) and these differences are a character of the bacteriophage: they are a reflection of the differences in the way in which the phages lyse their host bacteria. There is a strain, the phage T2, for example, which produces plaques on a lawn of *E. coli* that are small in size and have sharp edges. This is the normal form of this virus and the gene for this character is assigned the symbol r^+. A mutant strain is known which causes large plaques with fuzzy edges: this has the symbol r (= rapid lysis).

Another character is **host range**. Each phage strain can normally infect and lyse only certain strains of host bacteria, but mutants occur in which the host range is altered. The h^+ strains of phage T2, for instance, can lyse *E. coli* strain B but not B2, whereas the h mutant can lyse both B and B2. The best way to distinguish between the h^+ and the h strains is to grow them on a 'lawn' of bacteria which is a mixture of *E. coli* B and B2 (B/B2). On such a lawn the h^+ phages produce **turbid** (partially clear) plaques, because they can lyse only one of the two host strains, whereas the h types can lyse both of them and they produce clear plaques.

Recombination in bacteriophages

How can viruses exchange genes?

In order for recombination to take place between two virus particles, their DNA molecules must first come into physical contact. This can happen when two strains both infect the same bacterium, i.e. when there is a **mixed infection**. It can be arranged experimentally if a mixture of the two 'parental' viruses is added to a culture of the host bacterium in the proportions of 2 viruses:1 bacterium. On average each bacterial cell will then be infected by two viruses, and many of the pairs will be *mixed*.

Hershey and Rotman demonstrated recombination in the phage T2, using the visible mutants for plaque morphology and host range described above. Genotypes of the parental strains were r^+h (small clear plaques on B/B2) and $r\,h^+$ (large turbid plaques on B/B2). A mixed infection of the two strains was made in a culture of *E. coli* strain B (in which both strains could grow) and the lysate (products of lysis) plated out on a lawn of B/B2 to screen the progeny. In addition to the two parental types they also found two new plaque types—large, clear (*r h*) and small turbid (*r⁺h⁺*), which they concluded arose as a result of recombination between the two parental strains (Fig. 12.16).

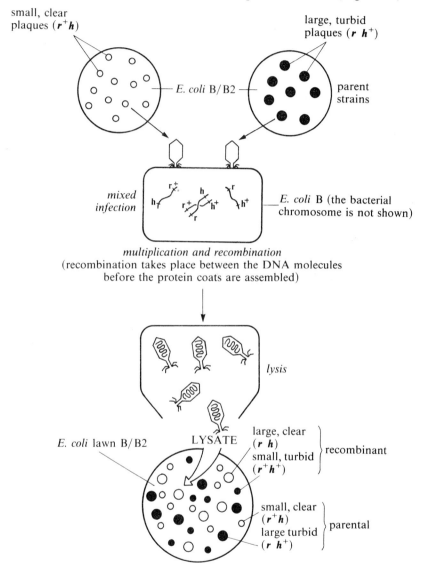

Fig. 12.16 Recombination in the bacteriophage T2.

Summary

The organisation of heredity and variation is different in prokaryotes and viruses from that in higher plants and animals (eukaryotes). Prokaryotes and viruses have no nuclei and no division processes of mitosis and meiosis. Their 'chromosomes' consist of a single molecule of 'naked' DNA (RNA in some cases) which is packaged within the cell or virus particle. Their relative simplicity, rapid reproduction and ease of culture on defined media, make them ideal for studying the biochemical and molecular aspects of genetics. As they only contain one 'chromosome' with only one linkage group, they have no independent segregation. Recombination takes place in a variety of ways. Bacteria can bring their chromosomes together for the exchange of genes by conjugating in pairs, by the entry of DNA fragments through the cell wall (transformation) or through the agency of bacteriophages (transduction). Viruses can pair up their chromosomes for recombination when two of them infect the same host cell.

Further reading

Bainbridge, B. W. (1980), *Genetics of Microbes*, Blackie.

Questions

1 What are bacteriophages and how has their study advanced our knowledge of inheritance?   *(O.L.E. Bot., 1981)*
2 What can eukaryotes do that prokaryotes cannot do? What can prokaryotes do that eukaryotes cannot do?   *(O.L.E. Bot., 1981).*

13
DNA and chromosomes

In Chapter 1 we stated that genetics is concerned with the transmission, the structure and the action of the information contained within the nucleus. Chapters 2 to 12 have been largely to do with *transmission*. We have covered all of the basic principles of genetics and discussed the way in which the inheritance of genes is related to the behaviour of chromosomes during division of the nucleus. We have also discussed some aspects of the way in which genes act, and interact with one another, in determining characters. In all of these discussions we have regarded genes, as Mendel did, simply as 'units of heredity'. Up until the 1950s this was the only view that geneticists could have, because before that time little was known about the structure and chemical composition of the genetic material. The laws, or principles, of heredity were all worked out without any firm knowledge of the nature of the genes or of the chromosomes which carry them. In the 1950s the chemical nature of heredity was discovered. The identification of the molecular structure of the genetic material led to a new understanding of the gene.

In this chapter we will present the evidence to show that the genetic material is made up of a chemical substance called deoxyribonucleic acid (DNA). We will deal with the molecular structure and replication of DNA molecules and show how they are organised within the chromosomes. In Chapter 14 we will explain how the information in DNA is utilised and what precisely we mean by 'the gene', in molecular terms.

Properties of the genetic material

At the outset it may be useful to outline the properties we would expect to find in a substance which serves as the material of heredity:

1. Obviously it must have capacity for **information storage**. Living organisms are highly complex in their structure and function and an enormous program of instructions must be present in the nucleus of the fertilised egg in order to direct their growth and development.
2. The genetic material must have the property of **self-replication**. It must be able to make identical copies of itself and to pass on the same program of instructions from one cell to another during growth and development, and from parent to offspring during reproduction.
3. For heredity to work, and for species to maintain their identity and continuity, the material of the genes must be **stable** in its structure.
4. There must also be **capacity for change**. Genes have to be stable, but at the same time they need to mutate to new forms—**alleles**. A low rate of error in their replication is essential as a source of genetic variation. Without some 'mistakes' in the copying process the genes would be fixed in one

form and there would be no possibility for species to change and to evolve over long periods of time.
5. In accordance with the chromosome theory of heredity the genetic material in eukaryotes must be **located mainly in the nucleus**—in the chromosomes.

The evidence that DNA is the genetic material

It was widely believed at one time that proteins were the most likely class of molecule to serve as the hereditary material, because they alone appeared to have enough diversity of structure to carry the genetic information. Proteins, however, are not self-replicating molecules. They are also unstable in the sense that they are continually being degraded and re-synthesised within the cell. Another possible candidate was nucleic acid, because it was found within the nucleus. It was first identified by Friedrich Miescher, in 1870. He isolated a **macromolecular** substance from pus cells (white blood cells) and salmon sperm, and called it 'nuclein'. It was later renamed as **deoxyribonucleic acid**, or **DNA**. In 1937 Feulgen demonstrated that most of the cell DNA was located in the nucleus.

The problem with DNA was that it appeared to have too simple a molecular structure. It was difficult for geneticists to see how it could fulfil all of the properties required of the genetic material, and for a long time therefore DNA was largely ignored. Eventually a number of lines of evidence, and some decisive experiments, confirmed that DNA was the genetic material. This evidence is summarised below.

Identification of the transforming substance   The first indication that DNA was the genetic material was provided by Avery, MacCleod and McCarty in 1944. They repeated and modified the transformation experiments of Griffith, with *Diplococcus pneumoniae*, which we described in Chapter 12. Griffith had shown that when mice were injected with a mixture of dead cells of a virulent S strain, and live cells of a non-virulent R strain of *D. pneumoniae*, some of the R cells were transformed into virulent S types by the presence of the dead S cell fragments (p. 159). Griffith had no idea which constituent of the bacterial cell acted as the transforming substance. Avery and his colleagues found this out. They systematically isolated and purified all of the major chemical constituents of the virulent bacteria, and then tested them for ability to transform when mixed with non-virulent R cells. They found that only the DNA fragments, and not proteins, were effective in causing transformation. In other words DNA fragments from one strain of bacteria were able to enter the cells of another strain and then bring about a change in its heredity.

Reproduction of bacterial viruses   The most conclusive and compelling evidence that DNA is the genetic material came from an ingenious experiment devised by Alfred Hershey and Martha Chase in 1952. They worked with the bacterial virus T2 which used *E. coli* as its host. As we have described in Chapter 12, bacteriophages reproduce by infecting host bacterial cells and then using the biochemical facilities of their hosts to manufacture more copies of themselves. Bacteriophages (phages) are small particles composed only of protein coats enclosing a molecule of nucleic acid (usually DNA) within the 'head' (Fig. 12.11). It is obvious that during the infection process the phages must introduce genetic material into the host cell to determine their own reproduction. Since the phage is such a relatively simple structure it provides the ideal situation with which to find out whether this material is the protein or the DNA. Hershey and Chase labelled the protein with radioactive sulphur (^{35}S) and the DNA with radioactive phosphorus (^{32}P), and were thus able to 'tag' the molecules during phage infection and reproduction. It was the DNA which entered the host cell and which was subsequently 'inherited' by the progeny viruses. Details of the experiment are given in Box 13.1.

Box 13.1   The Hershey–Chase experiment

To determine whether genetic information resided in the protein coats, or the DNA of the head of phage T2, Hershey and Chase used radioactive chemicals to label the two kinds of molecules. They grew some viruses in *E. coli* cells which were grown on a medium containing a radioactive isotope of phosphorus (^{32}P), and a separate batch in *E. coli* grown on a medium containing radioactive sulphur (^{35}S). Phosphorus is found in DNA, but not in protein, and sulphur occurs in protein but not in DNA. When the viruses were isolated from the bacteria one batch therefore had radioactive protein coats, containing ^{35}S, and the other batch had radioactive ^{32}P-labelled DNA in their 'heads'.

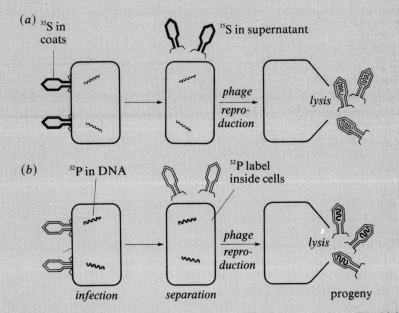

Fig. 13.1   The Hershey and Chase experiment with the bacteriophage T2 which showed that the DNA part of the virus, and not the protein, carries the genetic information for phage reproduction.

(*a*) Phages with ^{35}S-labelled protein were found to leave their radioactive 'ghosts' outside of the host bacterial cells, and the ^{35}S label was not inherited in the progeny released after multiplication and lysis.

(*b*) ^{32}P label in DNA entered the host cells and was inherited in the progeny (but its concentration was diluted during phage multiplication).

These two cultures of labelled phages were then used separately to infect normal samples of *E. coli*. After infection, samples were violently shaken in a kitchen blender to separate off the bacteria from whatever part of the virus had remained outside of the cells. Intact bacterial cells recovered after centrifugation were found to contain almost all of the ^{32}P label, and only a small amount of the ^{35}S. The supernatant solution which contained the parts of the viruses not involved in infection carried most of the ^{35}S label and almost no ^{32}P. Evidently the genetic material for phage reproduction is the DNA. This is the molecule which entered the host cells and which was transmitted to the progeny phages. The protein 'ghosts' remained outside the cells and were not inherited (Fig. 13.1).

Chromosomes contain DNA   Chromosomes contain DNA and they also carry the genes which determine heredity. There is a complete parallel between the way in which genes and chromosomes are distributed during nuclear division (Chapter 6). This parallel is embodied in the chromosome theory of heredity and strongly suggests that the chromosomes carry the genetic information. Chemical analysis shows that DNA is carried in the chromosomes.

Constancy of DNA content of nuclei   Diploid nuclei from somatic cells in any one species, at corresponding stages of the mitotic cycle, all contain the same quantity of DNA. Gametic nuclei have half the amount, as would be expected from genetic experiments.

Stability of DNA   In contrast to other cell constituents, DNA remains stable and intact as a large macromolecule. It is not metabolised.

Mutations are caused by UV light   The strongest mutagenic effect of ultra-violet light is at wavelengths between 250 and 270 μm. These are also the wavelengths at which DNA shows maximum absorption. This coincidence suggests that UV-induced mutations result from changes in DNA. Chemical mutagens also act by causing changes in DNA (Chapter 15).

Although the evidence that DNA is the genetic material became established beyond doubt, one problem still puzzled biologists.
    How could such a simple molecule fulfil all the properties of the genetic material?
    The answer was found in the structure of DNA.

The structure of DNA

Although the structure of DNA was unknown, the constituent building blocks had earlier been isolated and their structure determined from chemical analysis of purified DNA molecules.

Constituents of DNA

DNA can be broken down into three constituents: a pentose (or 5-carbon) sugar, organic bases and phosphoric acid. The sugar is **deoxyribose**:

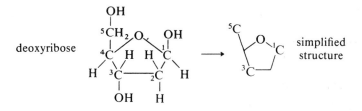

The organic bases are of four types: **adenine** (**A**), **guanine** (**G**), **thymine** (**T**) and **cytosine** (**C**). They are ring compounds which include carbon and nitrogen molecules. They fall into two groups—**purines** and **pyrimidines**—depending upon whether the ring is a double or single structure:

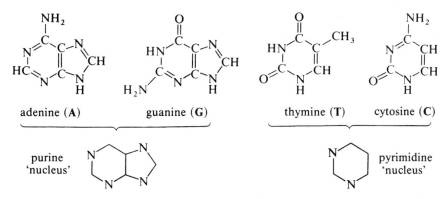

adenine (**A**)   guanine (**G**)   thymine (**T**)   cytosine (**C**)

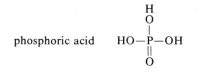

purine   pyrimidine
'nucleus'   'nucleus'

Phosphoric acid can be denoted as Ⓟ; it has the structural formula:

$$
\text{phosphoric acid} \qquad
\begin{array}{c}
H \\
O \\
| \\
HO-P-OH \\
\| \\
O
\end{array}
$$

The nucleotide

The basic unit of structure of the DNA molecule is the **nucleotide**. This structure contains one of each of the three kinds of constituent molecule—a base, a sugar and a phosphate group—as shown below (*a*). The base is attached to the sugar at carbon-1 (^1C). The phosphate group is linked to the ^5C atom. The nucleotide can be shown as a simple diagram (*b*) in which shapes are used to represent the constituent parts (S = sugar, P = phosphate, B = base):

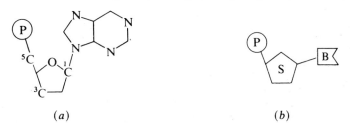

(*a*)   (*b*)

There are four different nucleotides corresponding to the four different bases: all of them have identical sugar and phosphate molecules. By a process of **polymerisation** nucleotides link together to form long chains called **polynucleotides**.

The polynucleotide

In the polymerisation process the linking of nucleotides is brought about by condensation reactions in which the phosphate groups bond alternately from position ^5C of one sugar molecule onto position ^3C of the next one, so forming a long -*sugar-phosphate-sugar-phosphate*- 'backbone'. When a polynucleotide chain 'grows' the nucleotides are added to the hydroxyl (OH) group at the ^3C position of the deoxyribose sugar. In other words they are always added at only one end of the chain, which is known as the 3' (3-prime) end. The other end of the chain is the 5' end. Because of the way in which the chains are formed they have a *direction*, with 'growth' at the 3' end (Fig. 13.2).

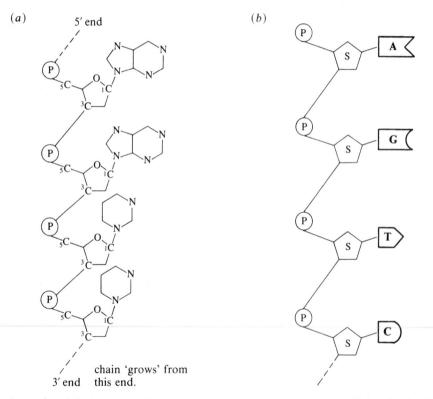

Fig. 13.2 (*a*) Simplified diagram of a small length of a polynucleotide chain showing how the individual nucleotides are linked together. The chain has a *direction* which is determined by the way the C atoms of the sugars are bonded to the phosphate groups. When a polynucleotide chain 'grows', nucleotides are always added to the 3' end.

(*b*) The polynucleotide may be drawn in an even simpler form by using shapes to represent the constituents. The large shapes of the four bases shown are the purines (**A, G**) and the smaller ones the pyrimidines (**T, C**).

The double helix

Once it became known for certain that DNA was the genetic material, and knowledge began to accumulate about the chemical nature of its constituent parts, many biologists were keen to unravel the secrets of its three-dimensional structure. In 1953 James Watson and Francis Crick, working in the Cavendish Laboratory in Cambridge, were the first to solve this puzzle. They showed how the polynucleotides were organised within the DNA molecule. A personal account of the events leading up to their discoveries is given in Watson's book '*The Double Helix*'. In this book Watson also refers to the major contributions made by other scientists—notably Maurice Wilkins and Rosalind Franklin of King's College London. In working out their model, Watson and Crick drew together knowledge from two main lines of enquiry: (1) chemical analysis and (2) X-ray crystallography.

Chemical analysis   In 1949 Erwin Chargaff had already made an important discovery about the base composition of the DNA extracted from a number of different species. Chargaff found that irrespective of its source the DNA always showed equivalent amounts of the purine base adenine (**A**) and the pyrimidine base thymine (**T**)—(**A** : **T** = 1 : 1). The same was true for guanine (**G**) and cytosine (**C**). In contrast there was no such fixed relation between the quantities of the two purines, **A** : **G** or the two pyrimidines, **T** : **C**. Nor was there any relation between **A** : **C** and **G** : **T**. These findings, which became known as **Chargaff's rules**, can be summarised as follows:

1. The number of purine bases (**A** + **G**) = the number of pyrimidine bases (**T** + **C**).
2. The number of adenine bases = the number of thymine bases (**A** : **T** = 1 : 1).
3. The number of guanine bases = the number of cytosine bases (**G** : **C** = 1 : 1).

Obviously these rules imposed some restrictions upon the way in which the bases could be arranged within the DNA molecule. Watson and Crick's interpretation of Chargaff's rules was that the base **A** was always paired with **T**, and **G** with **C**. This could only be achieved if DNA consisted of two strands held together by specific base pairing.

X-ray crystallography   Wilkins and Franklin had been working on X-ray diffraction patterns for some time. When X-rays are passed through crystalline preparations of DNA they are scattered in a certain way according to the arrangement of the atoms within the molecules. The scatter pattern can be recorded by allowing the X-rays to impinge on a photographic film. Spots on the film then reveal information about the angle of scatter and the underlying three-dimensional structure of the molecules in the crystal (Fig. 13.3). Identical patterns were obtained for DNA taken from T2 bacteriophage, various bacteria, the trout and the bull. These X-ray diffraction pictures showed that the DNA was a long thin molecule with a constant diameter of 2.0 nm and that it was coiled in the form of a helix. The helix made one full twist for every 3.4 nm of its length. As there was a distance of only 0.34 nm between each pair of bases there were reckoned to be ten bases per twist. The density of atoms in the molecule also indicated that it must be composed of *two* strands.

Having assembled all of this information, and having taken into account certain other relevant facts about the angles of bonds and the distances between

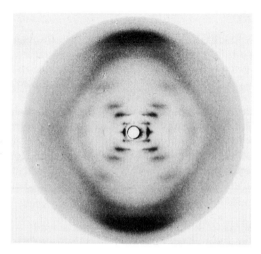

Fig. 13.3 The original X-ray diffraction photograph used by Watson and Crick in their work on the structure of DNA. The photograph was taken by Rosalind Franklin in Maurice Wilkin's laboratory. The helical structure of the molecule is indicated by the dark areas that form a cross in the centre of the photograph. (Original photograph by courtesy of Dr M. H. F. Wilkins, Biophysics Department, King's College London.)

atoms, Watson and Crick set about building a model—using pieces of wire and flat metal shapes to represent the bases. They came to the conclusion that the only way to fit all the constituents together was in the form of a *double* helix composed of two polynucleotide chains held together by hydrogen bonding between pairs of bases (Fig. 13.4).

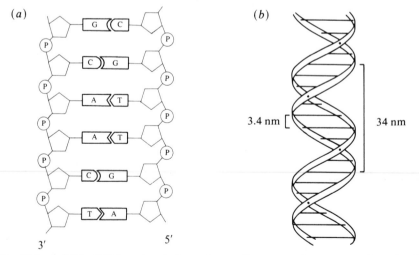

Fig. 13.4 (*a*) The DNA molecule is made up of two complementary polynucleotide strands which are anti-parallel (i.e. run in opposite directions). The sugar–phosphate backbone is on the outside of the molecule and the bases are on the inside. Hydrogen bonding between specific pairs of bases holds the two strands together.

(*b*) The strands are twisted into a double helix. Continuous lines represent the sugar–phosphate backbone and horizontal lines the base pairs.

The salient features of the model may be summarised as follows:

1. The DNA molecule is a **double helix** made up of two polynucleotide chains which are coiled around the same axis and interlocked. They can only be separated by untwisting—not by simply pulling apart sideways.
2. The sugar–phosphate backbones are on the outside of the molecule and the **bases are on the inside**.
3. The chains are held together by **hydrogen bonding between specific pairs of bases**, according to Chargaff's rules. Adenine only pairs with thymine (**A**:**T**) and guanine only pairs with cytosine (**G**:**C**). These particular combinations give the most efficient 'lock-and-key' arrangement of hydrogen bonding and are the *only pairs* which can fit together within the dimensions of the molecule (Fig. 13.5).
4. Because of the specific base pairing, the sequence of nucleotides in one strand determines the sequence in the other one. The two strands are therefore **complementary**.
5. The bases in the two strands can only be made to fit together if the sugar molecules to which they are attached point in opposite directions; i.e. the strands are **anti-parallel**.

The molecule can be thought of as a ladder in which the base pairs are the rungs and the sugar–phosphate backbones represent the two sides. The latter is then twisted into a double helix in which there are ten nucleotides per turn. The bases are flat structures arranged at right angles to the long axis of the molecule.

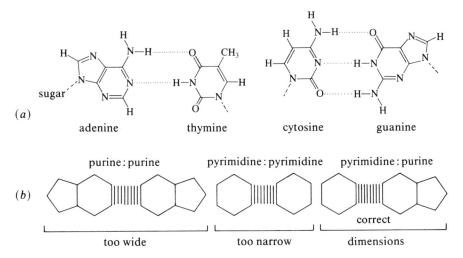

(a) adenine   thymine   cytosine   guanine

(b) purine:purine   pyrimidine:pyrimidine   pyrimidine:purine

correct

too wide   too narrow   dimensions

Fig. 13.5 (*a*) The two polynucleotide strands of DNA are held together by hydrogen bonds which form between *specific pairs of bases*. There are only two pairings which can fit within the space allowed in the molecule: these are the purine–pyrimidine pairs of **A**:**T** (2 bonds) and **C**:**G** (3 bonds). The other purine–pyrimidine combinations of A–C and G–T cannot bond together in the same way because they don't have a suitable arrangement of their atoms.

(*b*) Pyrimidine:pyrimidine and purine:purine pairs of bases have the wrong dimensions to fit within the double helix.

In 1962 Watson and Crick, together with Maurice Wilkins, were awarded the Nobel Prize for Medicine for their work on elucidating the structure of DNA.

Size of DNA molecules

In Fig. 13.4, only a tiny stretch of double helix is shown. In reality the molecules are extremely long and can be composed of several million nucleotides. It is now known that each chromatid of a eukaryotic chromosome contains one continuous molecule of DNA double helix running throughout its length, as explained later (p. 181). The size of a DNA molecule therefore depends on the size of the chromosome in which it is carried.

Implications of the Watson and Crick model of DNA structure

There are two important features of the Watson and Crick model in terms of the arrangement of the bases:

1. There are no restrictions upon the order in which the nucleotides can occur along one of the polynucleotide strands, and this suggests that the **sequence of bases** may be important as a means of storing and encoding the genetic information. This is precisely what Watson and Crick proposed. We will deal with this aspect in Chapter 14.
2. **Complementary base pairing** means that for any given sequence of the four bases in one of the strands, the sequence in the other one is determined. This suggested a mechanism by which DNA could self-replicate and make more identical copies of itself. The way in which this replication works is explained below.

Replication of DNA

When the two Cambridge scientists published an account of their model for DNA structure they also put forward a theory for the way in which it could undergo **replication**. According to this idea the two strands could unwind from one another, following the disruption of their hydrogen bonds, and each one could then serve as a **template** for the synthesis of a new complementary strand. The molecule could 'unzip' its bonds from one end, and then new bases present in the nucleus could be assembled alongside their complementary partners, in the correct sequence, and become linked up by the formation of a new sugar-phosphate backbone. When the unwinding and synthesis had passed along the full length of the molecule, two identical daughter molecules would result, each of which would be an exact copy of the original double helix. This was later called **semi-conservative replication**, because of the way in which the newly formed daughter molecules each contained (or conserved) one of the original 'parental' strands, plus one newly synthesised strand. The scheme is illustrated in Fig. 13.6. Later on this model was shown to be correct (complications in the replication of DNA, which arise because the two strands are made in opposite directions, are not discussed in this book).

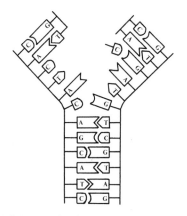

Fig. 13.6 Watson and Crick's model for the replication of DNA. The two parent strands are 'unzipped' and, once separated, each one then acts as a template for the synthesis of a new strand by complementary base pairing. As a result the original molecule is duplicated as two identical copies, each of which contains one old and one new strand.

From the bacterium *E. coli*, Kornberg (1957) purified the enzyme DNA-polymerase, which is responsible for the 'zipping up' of bases during the synthesis of new polynucleotide chains. He was then able to use this enzyme to direct the synthesis of new molecules of DNA in a test-tube. Single-stranded DNA templates, of known base composition, were mixed together with the polymerase enzyme and a supply of free nucleotides. Double-stranded DNA was then formed in which the base sequence of the new strand was complementary to that of the existing polynucleotide. This experiment showed that the replication of DNA was indeed based upon specific pairing between complementary bases; but it didn't prove that the process was *semi-conservative*. This was demonstrated by an experiment of Meselson and Stahl carried out in 1958: the details are given in Box 13.2.

Box 13.2   The Meselson–Stahl experiment

Meselson and Stahl worked with the bacterium *E. coli* because it has a chromosome composed of only one molecule of DNA, and this replicates each time the cell divides (Chapter 12). In a synchronous culture of cells (i.e. all cells at the same stage of their reproductive cycle) the DNA undergoes a round of replication each time the number of cells is doubled (each **generation**). By monitoring the doubling time of a culture it is therefore possible to follow the cycles of DNA replication.

When DNA molecules are extracted from *E. coli* cells and centrifuged in a solution of caesium chloride, they migrate to form a distinct band at a position in the tube where their density matches that of the salt molecules. The salt itself forms a gradient of increasing density from the top to the bottom of the centrifuge tube. What Meselson and Stahl did was to grow their *E. coli* for several generations on a medium containing the heavy isotope of nitrogen (^{15}N) as the only nitrogen source. Nitrogen atoms are found in all the DNA bases and when the DNA from their culture was extracted and placed in a caesium chloride density gradient it settled as a *heavy* band. This band was much lower down the tube than the normal (light) DNA which has ^{14}N in its bases. In other words they labelled the DNA of the chromosomes and made it heavy. These ^{15}N-labelled bacteria were then transferred to a new medium containing ^{14}N as the only nitrogen source, and were allowed to grow and to duplicate their chromosomes over several generations. After each of three successive generations a sample of the cells was removed and the density of their DNA determined. The first sample revealed the presence of DNA with intermediate density. This must correspond to a 'hybrid molecule' containing one heavy strand and one light strand. In the second sample two types of DNA were present in equal proportions—hybrid DNA and light DNA. In the sample taken after the third generation, and therefore the third cycle of replication, three-quarters of the molecules were light and one quarter were of hybrid, intermediate density. The results, together with their interpretation, are shown in Fig. 13.7. They can only be accounted for on the basis of *semi-conservative replication*: and they confirmed the Watson–Crick model.

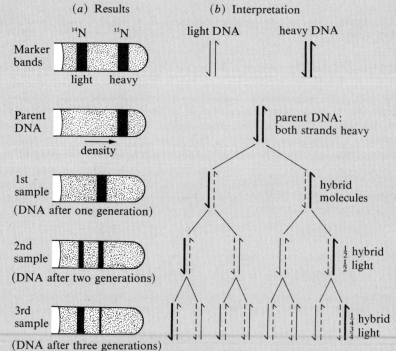

Fig. 13.7 Meselson and Stahl's experiment which provided conclusive evidence for the semi-conservative replication of DNA.

(*a*) The diagram represents the position of DNA molecules in the caesium chloride density gradients.

(*b*) Interpretation of the pattern of replication. Polynucleotide DNA strands are represented as arrows. Dashed arrows are the newly formed strands.

The results in (*a*) can only be explained if DNA replicates semi-conservatively, such that each strand of an existing double helix serves as a template for the synthesis of a new strand—as shown in (*b*).

DNA and chromosome structure

The chemical composition and molecular structure of DNA are normally studied using highly purified samples of nucleic acid extracted from cells and nuclei. In its *natural* form, of course, DNA is not a highly purified crystalline substance. It is a component of the chromosomes. We have already discussed chromosomes (Chapter 2) in relation to the distribution of genetic information during division of the nucleus in mitosis and meiosis. We must now look into their molecular organisation, and ask the question: how is DNA organised in the chromosomes?

There are in fact two kinds of chromosomes: the simple naked ones found in viruses, bacteria and cell organelles, and the more complex ones which occur in the nuclei in eukaryotes. We will be concerned principally with the latter.

Prokaryotic chromosomes

As we have described in Chapter 12, the chromosomes of prokaryotes are simply long naked molecules of double-stranded DNA. The bacterium *E. coli*, for example, has a single circular chromosome, about 1.4 mm in length, which contains four million nucleotide pairs. It replicates in a semi-conservative fashion, beginning at a fixed point and proceeding in one direction around the length of the chromosome at a rate of 16 000 base pairs per minute. Relatively simple chromosomes of this type are also found in many DNA viruses, as well as in chloroplasts and mitochondria of higher organisms (Chapter 9).

Eukaryotic chromosomes

In eukaryotes the nucleus of each cell contains several pairs of chromosomes. The chromosomes are large structures and they change their form and structural organisation at different stages throughout the cell cycle (Chapter 2). They are composed of about equal amounts of both protein and DNA. The DNA, of course, carries the genetic information and the proteins are mainly concerned with structure. The chemical complex of DNA and protein which is found in chromosomes is known as **nucleoprotein** or **chromatin**. The proteins are of two types—basic **histones** and **acidic proteins** (non-histones). There is a large variety of non-histones, but only five types of basic histones (H1, H2A, H2B, H3 and H4) which are common to most species.

Eukaryotic chromosomes contain an enormous quantity of DNA. In *Drosophila*, for example, the largest chromosome of the complement has 62 000 000 nucleotide base pairs in each chromatid. The largest one in man has more than 200 000 000. At one time it was widely believed that each chromatid must be multi-stranded, or composed of several helices of DNA running through it in parallel. This view is no longer held, except in special cases such as the giant polytene salivary gland chromosomes of *Drosophila* and other dipteran flies (see Chapter 17). We now know that each chromatid contains *one* uninterrupted double helix running throughout the length. This was proved by Kavenoff and Zimm in 1973 when they managed to isolate intact DNA molecules from *Drosophila* chromosomes. The amount of DNA in a chromatid of the largest chromosome was completely accounted for by the *one* molecule of double helix that came from it.

The sheer quantity of DNA in a eukaryotic chromosome poses a problem of organisation. This organisation must allow for the changes which take place during cell division. At interphase the DNA needs to extend itself in order to replicate and allow for the activity of genes; while in nuclear division it must become tightly packed and shortened in order for the chromosomes to move and to divide during mitosis and meiosis. Consequently, the DNA is not present in a fully extended form in the chromosome. It is 'packaged' in some way so that 7 cm of DNA helix which is found in chromosome 1 of man, for instance, is all accommodated within a metaphase chromosome which is only 10 μm long. In this case the **packing ratio** of extended DNA : metaphase chromosome, is of the order of 10 000 : 1. At interphase, when the chromosomes are much less tightly coiled, the packing is about 100 : 1.

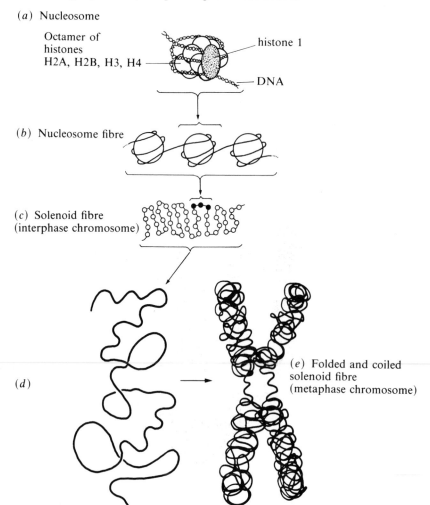

(*a*) Nucleosome

Octamer of histones H2A, H2B, H3, H4

histone 1

DNA

(*b*) Nucleosome fibre

(*c*) Solenoid fibre (interphase chromosome)

(*d*)

(*e*) Folded and coiled solenoid fibre (metaphase chromosome)

Fig. 13.8 DNA and chromosome organisation.

Organisation of DNA in chromosomes   When chromatin from an interphase nucleus is spread out and viewed under the electron microscope it has a definite repeating structure which resembles a string of beads. The 'beads' are **nucleosomes**. They consist of an **octamer** of two molecules each of histones H2A, H2B, H3 and H4, around which is wrapped $1\frac{1}{3}$ turns of DNA (Fig. 13.8a). There are 146 nucleotide pairs within the DNA encircling the nucleosome, and when the length of double helix which runs to the next nucleosome is included, the total length is about 200 nucleotides. The nucleosome is the basic unit of structure of chromatin. The nucleosome fibre shown in Fig. 13.8b is what we see when interphase chromatin is spread out for analysis; in its natural state, the chromatin fibre (solenoid fibre) is three times thicker (Fig. 13.8c,d). Packing of nucleosomes into the solenoid fibre is thought to be the main function of histone H1 which is attached to the outside of the nucleosome.

The turns that the DNA makes about the nucleosome give a packing ratio of 10:1, and with further coiling into the solenoid fibre a ratio of 100:1 is achieved. When the chromatin is condensed (by coiling and folding) during cell division, into a metaphase chromosome, a further packing occurs to bring the ratio up to 10 000:1. Details of this final level of organisation are unknown, but it is thought that the acidic (non-histone) proteins form a 'scaffold' around which chromatin fibres are looped. Chromatin fibres are clearly visible in the photograph of a metaphase chromosome shown in Fig. 13.9.

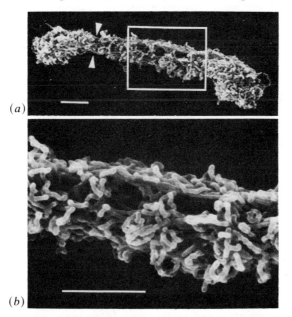

(a)

(b)

Fig. 13.9   (a) A single human chromosome at metaphase of mitosis photographed on the scanning electron microscope. Arrows indicate the position of the centromere (individual chromatids not clearly visible).

(b) Higher magnification of the framed section (bars represent 0.5 μm). The photographs show the complexity of the metaphase chromosome and reveal some details of its organisation at the level of the *chromosome fibre*. See text for further details.

(Reproduced with permission from K. R. Utsumi, 1982, *Chromosoma* **86**, 683–702.)

Replication of DNA in chromosomes   Since the chromatid contains one double helix of DNA it is able to replicate semi-conservatively, in the same way that naked DNA molecules do, during the synthesis (S) phase of the mitotic cycle. Because there is so much DNA in a chromosome, however, the process is rather more complicated, as explained in Box 13.3.

Box 13.3   Replication of DNA in chromosomes

DNA molecules in eukaryote chromosomes are too long to replicate their whole length from only one initiation point—as they do in prokaryotes (p. 154). It is impossible to untwist all of the DNA in a chromatid from one end through to the other. In *Drosophila* the replication fork moves at a rate of 2600 bases/minute. At this rate it would take more than 16 days for the longest chromosome to duplicate itself if it had only one origin of replication. In fact it replicates in less than three minutes, by opening up more than 6000 replication forks at different sites along the chromosome. A short stretch of DNA with two replication sites is shown below (Fig. 13.10)—the dashed lines are newly synthesised polynucleotide strands:

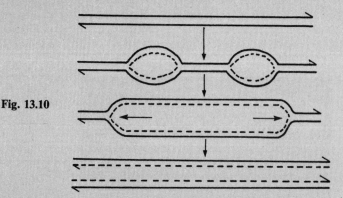

Fig. 13.10

At each site the replication forks move out in two directions so that eventually they all meet up with one another. When replication begins at a site it is necessary for one of the DNA strands to break, in order for the two strands to unwind and allow synthesis of the new ones by complementary base pairing. There are enzymes which can cause single-stranded 'nicks' in DNA and others which can join them back together again (such enzymes are also needed for the breakage and rejoining of DNA strands which takes place during crossing over at meiosis (Chapter 2)). During DNA replication in the interphase chromosome, the existing nucleosomes remain associated with one of the newly formed chromatids. New nucleosomes are then made to organise the nucleoprotein complex of the other chromatid.

DNA, chromosome structure and replication   In essence each chromatid contains one helix of DNA running throughout its length. Replication of the chromatid is semi-conservative and follows the same pattern as the DNA which it carries. We can therefore represent the mitotic cycle of chromosome duplication and division in terms of the DNA helix, as given in Fig. 13.11. This is the simplest possible relation between DNA and chromosomes, and overlooks the complexities of organisation that were discussed above.

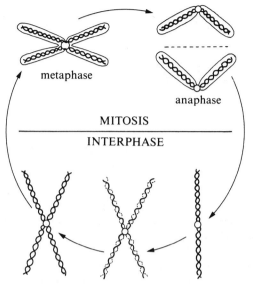

metaphase

anaphase

MITOSIS

INTERPHASE

DNA synthesis and chromosome duplication

Fig. 13.11 Simplified diagram showing the way in which DNA is organised in the structure and replication of the chromosome.

Summary

The genetic material which carries the information for heredity is a chemical substance called deoxyribonucleic acid (DNA). It is a long molecule made up of two chains of polynucleotides which are twisted together into a double helix. The polynucleotides are composed of four kinds of organic bases—adenine, thymine, cytosine and guanine—which are linked together in a long chain by their attachment to sugar–phosphate backbones. Hydrogen bonding between specific pairs of bases holds the polynucleotides together and also provides the mechanism for their self-replication. The sequence of bases along a polynucleotide encodes the genetic information. In prokaryotes, viruses and cell organelles, the chromosomes are simply naked molecules of DNA. Eukaryote chromosomes are composed of nucleoprotein fibres which are a complex of DNA and protein. The organisation of eukaryote chromosomes is complicated but in essence each chromatid consists of one molecule of double helix running throughout its length.

Further reading

Watson, J. D. (1970), *The Double Helix*, Penguin Books.
Watson, J. D. and Crick, F. H. C. (1953), A structure for deoxyribose nucleic acid, *Nature* **171**, 737–738.

Questions follow Chapter 14.

14
DNA and genes

It is evident from what was written in the previous chapter that DNA carries the genetic information. It is evident too that one of its functions is to make more identical copies of itself, so that heredity can work and exact copies of the genetic blueprint can be handed on, when cells divide and when organisms reproduce. DNA makes more DNA, and the way in which it is organised within the chromosomes ensures that it is regularly distributed during the nuclear divisions of mitosis and meiosis.

DNA must also act. It contains the genes and the information it carries must be utilised in some way to determine characters, and to direct the growth, development and reproduction of living organisms.

In this chapter we will discuss the way in which the genetic information is encoded in the DNA. We will also describe the way in which it is utilised within cells and what precisely we mean by the idea of a 'gene'. Before we can discuss any of these things, however, we must first of all establish the link between genes and proteins.

Genes and proteins

It comes as a surprise to learn that the first idea about how genes could act was put forward as early as 1909 by the English physician Sir Archibald Garrod. At that time the gene was a new *concept*, and nobody had the faintest notion about its chemical composition or structure. Garrod thought that genes may exert their effects through enzymes. He suggested that a rare human disease, called **alkaptonuria**, was due to a block in the metabolism of the amino acid tyrosine. The disease was inherited as a simple Mendelian recessive character, and Garrod surmised that the "inborn error of metabolism" came about because of the failure of the recessive allele (in homozygotes) to control the synthesis of an enzyme needed for the metabolism of tyrosine. The disease, although benign (harmless), is very distinctive. Because the enzyme homogentisate oxidase is lacking, homogentisic acid accumulates (Fig. 14.2) and is excreted in the urine. After standing the urine turns black.

A few years later, a much more serious metabolic disorder known as **phenylketonuria** (PKU) was discovered. This disease causes severe mental retardation and is due to an absence, or deficiency, of the enzyme phenylalanine hydroxylase. The amino acid phenylalanine cannot be converted into tyrosine (Fig. 14.2) and it accumulates in all of the body fluids with drastic consequence. Therapy for the condition is a diet low in phenylalanine. It is inherited as a simple recessive character determined by one gene, and again there is a link between the gene mutation and a defect, or absence, of an enzyme.

The suggestion that genes work through their control over the production of enzymes was therefore born very early on in the history of genetics. It came to the forefront of our thinking, however, in 1941 when Beadle and Tatum published the results of their pioneering work on the biochemical genetics of the bread mould *Neurospora crassa*. They were able to show beyond doubt that a mutation in a single gene resulted in a change in the activity of a single enzyme.

One gene, one enzyme

Neurospora is an ideal organism with which to study the way that genes determine characters. It is an ascomycete fungus with a life cycle very similar to that described for *Sordaria* in Chapter 3. Since it is haploid for the major part of its life cycle, recessive mutations are expressed in the phenotype—and are readily detected. Crosses between strains are easily made (as in *Sordaria*) by growing opposite mating types together in the same culture. The sexual cycle is completed in as little as ten days and results in fruiting bodies (perithecia) full of sac-like asci. Each ascus contains eight large haploid ascospores which can be dissected out and then grown in order to test their genotype.

Wild-type *Neurospora* can be cultured in the laboratory on minimal agar medium. The only requirements are water, sucrose as a source of carbon, and mineral salts—including nitrates, sulphates and phosphates. The fungus has a full complement of enzymes which enable it to use these simple ingredients to manufacture all of the other essential substances needed for normal growth and development. Mutants of the wild strain can be obtained that are unable to make certain substances, such as a particular amino acid or vitamin, required for their nutrition. As described in Chapter 12 such nutritionally defective strains are known as **auxotrophic mutants**, in contrast to the normal **prototrophic** strain. Auxotrophs will only grow on a minimal medium when it is supplemented with the particular substance that they are unable to make for themselves. Alternatively, they can be grown on a complete medium.

Beadle and Tatum set out to isolate auxotrophic mutants of *Neurospora*, in order to use the simple biochemical defects as *characters*. The procedure they used is illustrated in Fig. 14.1. A number of different mutants were obtained, each having a different specific growth requirement. They were kept as pure cultures and studied biochemically. In each case the mutant was unable to grow because it could not synthesise a particular amino acid or vitamin. It was argued that this defect was due to the lack of a single enzyme which was needed to complete the relevant biochemical pathway. This argument was supported by the fact that the mutants accumulated certain substances which were normally found in only small amounts. This would happen if the pathway was blocked and the substance made in the step before the block was accumulating. The character which distinguished each mutant from the normal wild strain was therefore a simple biochemical one, involving the absence or presence of a single enzyme involved in the nutrition of the fungus.

When crosses were made between each of the mutants and the normal wild strain the character difference was inherited as a single gene with two alleles. Ascospores taken out of hybrid asci gave four normal cultures and four mutants when grown and tested. In other words, a mutation in a single gene caused the absence, or loss of activity, of a specific enzyme.

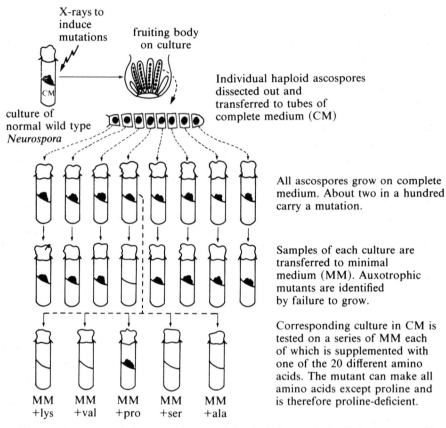

Fig. 14.1 Summary of the procedure used for isolating and classifying auxotrophic mutants in *Neurospora*. The scheme shows how a mutant which is unable to make a particular amino acid is detected. The same procedure is used for isolating mutants which are defective in the synthesis of other essential substances, such as vitamins. (Based on Beadle, 1948.)

These findings led Beadle and Tatum to formulate the hypothesis of **one gene, one enzyme**. They envisaged that all biochemical processes are under genetic control; that these processes proceed through a series of stepwise reactions; and that each step is catalysed by a single enzyme and *each* enzyme is controlled by *one* gene. This vision has proved to be essentially correct, except for some minor modifications which are described below. Figure 14.2 shows a small part of a biochemical pathway in man: each specific enzyme failure is associated with a well known human genetic disorder which is inherited as a simple recessive character. It illustrates the way in which genes direct the biochemical reactions of cells, through their control over enzymes, and thereby determine the characters of all living organisms.

In order to understand how genes determine enzymes we now need to know something about enzymes.

Enzymes belong to a class of substances called proteins. They catalyse the biochemical reactions that take place within cells. Each enzyme has a precise

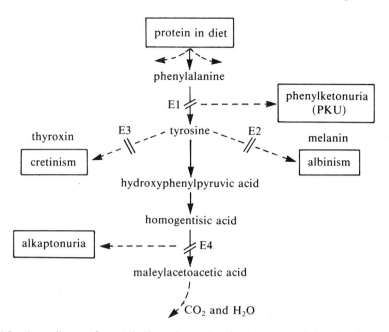

Fig. 14.2   A small part of a metabolic pathway showing how genes determine characters in man. Each block (─╫─) in the pathway is due to a defect in an enzyme (E). Each enzyme defect results in a genetic disorder which is known to be inherited as a simple recessive character.

structure which is necessary for its function. It seems a reasonable hypothesis, therefore, that genes might act by determining the *structure* of enzymes. Before we can examine this hypothesis we need to describe the structure of enzymes, i.e. the structure of proteins.

Protein structure

A **protein** is *an organic compound composed of amino acids which are linked together by peptide bonds to form long chains, called polypeptides.* There are 20 different **amino acid** subunits commonly found in proteins (Box 14.1). They can all be represented by the general formula:

$$H-N-C-C-OH$$

All amino acids have a carboxyl group (COOH) and an amino group (NH_2), which are attached to the same carbon atom. They differ only in the side-chain represented as ℝ. Amino acids link together by a condensation reaction between their carboxyl and amino groups, to form a peptide bond. Molecules of water are released in the process (see over).

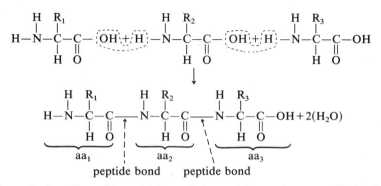

The long chains of peptides which result from this reaction are called **polypep-tides**. Each polypeptide has a free amino group at one end (amino, or N-terminus) and a free carboxyl group at the other end (carboxyl, or C-terminus). Different proteins differ in their amino acid composition and sequence.

Box 14.1 The 20 amino acids commonly found in proteins

Amino acid	Abbreviation	Examples of structure	
Alanine	Ala	$H_2N-\underset{\underset{H}{\vert}}{\overset{\overset{H}{\vert}}{C}}-COOH$	**Glycine** is the simplest amino acid.
Arginine	Arg		
Asparagine	Asn		
Aspartic acid	Asp		
Cysteine	Cys		
Glutamine	Gln		
Glutamic acid	Glu		**Tryptophan** has an aromatic side chain.
Glycine	Gly		
Histidine	His		
Isoleucine	Ile		
Leucine	Leu		
Lysine	Lys		
Methionine	Met		
Phenylalanine	Phe		**Methionine** has a sulphur-containing side chain.
Proline	Pro		
Serine	Ser		
Threonine	Thr		
Tryptophan	Trp		
Tyrosine	Tyr		
Valine	Val		

As there is virtually an infinite number of ways in which the 20 amino acids can be assembled to make up a polypeptide, there are a limitless number of proteins which may result. This is not to say, however, that proteins are made of random combinations of amino acids: on the contrary each one has a **specific sequence** that is essential to its structure and function. The example below shows part of the amino acid sequence of a gluten protein which gives wheat flour its characteristic bread-making quality. This protein contains about 500 amino acid residues, of which only 26 at the N-terminus and four at the C-terminus have been identified (Fig. 14.3).

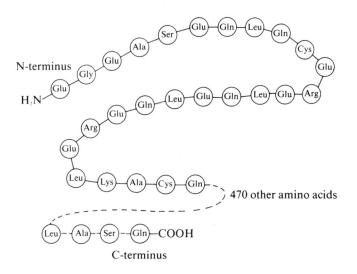

Fig. 14.3   (Courtesy of Peter Shewry, Rothamsted.)

To determine the sequence of amino acids in a protein (called 'sequencing'), chemical reactions are used to remove amino acids from the N-terminal end of the polypeptides, one at a time, and enzymes are used to take amino acids off the C-terminal end. For technical reasons this is done by first breaking up the molecule, and then sequencing the small fragments individually. The 'breaking' is done with proteolytic enzymes that **cleave** polypeptides at points where certain amino acids occur (e.g. trypsin which breaks the bonds next to arginine and lysine residues). Frederick Sanger sequenced the first protein in 1953, namely the hormone **insulin**, and showed that it had a precisely defined amino acid sequence. Complete amino acid sequences of several hundred different proteins are now known.

The specific sequence of amino acids in a polypeptide chain gives the protein molecule its **primary structure**. Amino acids which are close to one another in the chain then interact in various ways. This may cause, for example, one part of the linear polypeptide to coil up into a α-helix, and another part to form a β-pleated sheet. These interactions give the **secondary structure**. Disulphide bonds between sulphur-containing amino acids which are far apart in the chain stabilise the molecule in its characteristic three-dimensional, or **tertiary structure**. Proteins that contain more than one polypeptide have a further level of organisation known as the **quaternary structure**. Because these higher levels of organisation are determined by the position that various amino acids occupy in the primary chain, the correct sequence is vital if the protein is to perform its proper function. This is especially important in the case of enzymes which have an **active site**. The loss, or incorrect order, of amino acids will lead to a corresponding loss, or reduction, in enzyme activity.

The one gene, one enzyme hypothesis might therefore be explained on the basis that **genes determine the amino acid sequence of enzymes**.

One gene, one polypeptide chain

As we have already explained, enzymes are only one class of proteins. There are many other kinds which also play important roles in structure and metabolism.

Structural proteins such as collagen (in cartilage), keratin (hair and hooves) and actin and myosin (muscles) are vital components of the structure of animal bodies. **Storage proteins** act as reservoirs of food storage in the endosperm of seeds. **Antibody proteins** play a vital role in the body's defence mechanism against disease. Certain **hormones** (e.g. insulin) are also proteins: they take part in regulating the physiology and development of organisms. **Haemoglobin** is a **transport protein** involved in carrying oxygen in the blood. **Histones** and **tubulin** are important proteins concerned, respectively, with the structure of chromosomes, and with the spindle apparatus on which the chromosomes move during mitosis and meiosis.

It soon became clear that the structure of all of these proteins, and not just the enzymes, is specified by information encoded in the genes of the DNA. It became necessary therefore to make the one gene, one enzyme hypothesis more general, and to rephrase it as **one gene, one protein**. When it was further realised that the quaternary structure of proteins may be due to two (or more) different polypeptides, and that these could be determined by quite separate genes, it was necessary to modify the hypothesis again, and to phrase the modern statement of Beadle and Tatum's principle as **one gene, one polypeptide chain**. These modifications came about largely as a result of certain detailed studies on the structure of haemoglobin.

Haemoglobin is not an enzyme: it is a transport protein which carries oxygen. It is a substance which is found in the erythrocytes of the blood and which gives them their red colour. In normal adults the molecule is composed of *four* polypeptides—two identical alpha (α) chains and two identical beta (β) chains. These four chains associate together to give the quaternary structure of the active protein (Fig. 14.4). The α-polypeptides each have an identical sequence of 141 amino acids and the β-polypeptides 146. The two polypeptides are specified by two different genes which are unlinked. An abnormal form of haemoglobin in some people causes them to suffer from the blood disorder called **sickle-cell anaemia**. The disease is genetically determined and inherited as a simple Mendelian character. It is expressed in recessive homozygotes and results in a malformation of the erythrocytes which take on a 'sickle' shape (Chapter 20, Fig. 20.4). In 1956 Vernon Ingram used a new 'fingerprinting' technique for protein analysis, and showed that the difference between normal haemoglobin (HbA) and the abnormal form (HbS) was due to a difference of only *one* amino acid. **Valine** had been substituted for **glutamic acid** at position 6 near the N-terminal end of the β chains (Fig. 14.4).

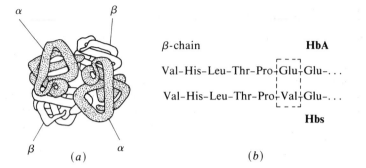

Fig. 14.4 Diagram showing (*a*) the two α and two β polypeptides which comprise the quaternary structure of haemoglobin and (*b*) the single amino acid substitution in the chain which changes **HbA → HbS** (the diagram shows only 7 amino acids at the N-terminal end of the β chain).

Ingram's research clinched the idea that genes bring about their effects on characters by controlling the structure of proteins. One gene provides the information for one polypeptide chain. Mutations alter the phenotype because they affect the amino acid sequence of the proteins and change the way that the proteins carry out their function within the cell.

How can this information be carried in the DNA?

The genetic code

Once it was established that genes worked by specifying the amino acid sequence of polypeptide chains, it was obvious that their information must be carried in the sequence of bases in the DNA. *The way in which the genetic information is encoded in DNA* is referred to as **the genetic code**.

On purely theoretical grounds there are various suggestions that can be made concerning the organisation of the genetic code. We can rule out straight away the idea that one base codes for one amino acid, because we have only four kinds of bases and there are 20 different amino acids. If bases are taken two at a time, along the sequence, then there are 16 (i.e. 4^2) possible combinations—taking the order into account, e.g. **AT** and **TA** are different. This is still insufficient to meet the requirements. When bases are taken three at a time however, as triplets, we see that there are 64 (4^3) different ways in which they can be combined together. This gives us more than enough bits of information to code for 20 amino acids. In fact we have 44 triplets to spare (see p. 201). Based on such theoretical arguments the **base triplet hypothesis** was put forward and the triplets became known as **codons**. It would be inefficient, of course, to have a code based on any more than the minimum number of three bases per codon. Experimental evidence in support of the base triplet hypothesis was provided by Francis Crick in the early 1960s. The details of his work are beyond the scope of this text, but some reference to the kind of experiments that he carried out is given in Chapter 15 (p. 209).

The question we must now come to consider is how the coded information contained in the DNA is utilised in the cell to determine the structure of proteins.

Utilisation of genetic information

In eukaryotic organisms the DNA is located in the chromosomes, within the nucleus, while the assembly of amino acids into proteins takes place in the cytoplasm and outside of the nucleus. Information therefore has to be transferred from the site where it is stored to the site where it is utilised within the cell. Both the transfer and the utilisation are accomplished through the agency of another kind of nucleic acid known as **ribonucleic acid**, or simply **RNA**. RNA molecules are found in large quantities in cells and tissues that are actively engaged in the manufacture of proteins.

Ribonucleic acid (RNA)

RNA is a nucleic acid and is therefore similar in both its chemical composition and its structure to DNA. It is composed of nucleotides which are polymerised into chains of polynucleotides. The main features which distinguish RNA from DNA are:

1. RNA is a single-stranded molecule: but it may be folded into various complex forms in which some double-stranded regions are found.

2. The sugar molecules in RNA are ribose, rather than deoxyribose as in DNA. In ribose sugar a hydroxyl group (OH) replaces the hydrogen which is present at the ^2C position in deoxyribose:

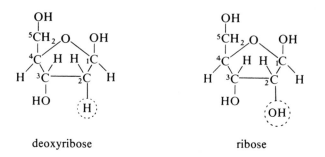

deoxyribose   ribose

3. In RNA the base uracil (**U**) replaces the thymine in DNA (but no one really knows why).

uracil   thymine

There are three different types of RNA molecules present in the cell, each of which plays a key role in the biosynthesis of proteins: these are messenger RNA, transfer RNA and ribosomal RNA.

Messenger RNA (mRNA), as its name implies, is the RNA that carries the genetic message between the DNA in the nucleus and the site of protein synthesis in the cytoplasm. In order for the message to be read off the DNA, the double helix first of all unwinds in the region of the gene, or genes, which are being expressed, and one of the strands then serves as a template for the synthesis of a complementary strand of mRNA. *The process in which mRNA is synthesised from a DNA template is known as* **transcription** (copying). Transcription is catalysed by a key enzyme called **RNA polymerase**. As the synthesis of the mRNA polynucleotide proceeds it is unzipped from the DNA template, and when it is completed the mRNA is transported across the nuclear membrane and into the cytoplasm. Figure 14.5 is a diagram of transcription.

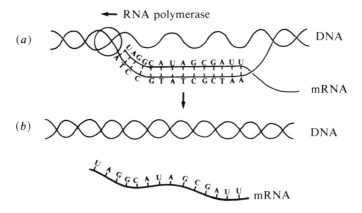

Fig. 14.5   Transcription of mRNA from DNA. The DNA helix unwinds for part of its length, from which the message is being taken, and one of the strands (the coding strand) acts as a template for the synthesis of a complementary single strand of messenger RNA. The process is catalysed by the enzyme RNA-polymerase, and in the formation of the polynucleotide strand uracil (**U**) in RNA pairs specifically with the base adenine (**A**) in DNA (*a*). After transcription is completed the DNA returns to its double-stranded form (*b*).

Transfer RNA (tRNA) is found in the cytoplasm. Its function is to pick up amino acids and to carry them to the sites of protein synthesis, on the ribosomes, so that they can be joined together into polypeptides. There are at least 20 different tRNA molecules, one for each amino acid, and they are all similar in their basic structure. They consist of a single strand of RNA, about 80 bases long, which is folded back upon itself in a clover-leaf arrangement due to pairing between complementary bases (Fig. 14.6). The tRNA molecules also contain some unusual nucleotides, such as pseudouridine and inosine, which are unable to form hydrogen bonds with other bases; they are found in the unpaired loops within the molecule.

One unpaired end to each tRNA contains a triplet of exposed nucleotides, known as the **anticodon**, which is complementary to one (or more) of the codons carried in mRNA. The other unpaired end has a site for attachment to a specific amino acid. Each tRNA therefore picks up its own amino acid, and by matching of its anticodon with the complementary codon in mRNA the amino acids can be assembled in the correct sequence—as explained below.

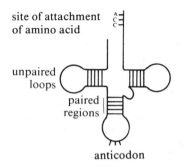

Fig. 14.6 Simplified diagram of the structure of transfer RNA (tRNA). Each tRNA molecule is characterised by having its own anticodon and a site of attachment for a specific amino acid.

Ribosomal RNA (rRNA) is a component of the **ribosomes**. These are the structures within the cell where the synthesis of proteins takes place. The ribosomes are often associated with the endoplasmic reticulum. They are uniform spherical structures composed of two parts—the small and the large subunits (Fig. 14.7). Each subunit is made up of about equal parts of RNA and of protein. The small subunit has one RNA molecule of approximately 1500 bases in length, and the large subunit has two RNAs—one of 3000 bases and one of 100 bases. The ribosomes of prokaryotes are a little smaller than those of eukaryotes, and in both groups there are several thousand of them present in each cell.

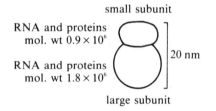

Fig. 14.7 Diagrammatic representation of a ribosome of *E. coli.*

Protein synthesis

Protein synthesis is *the linking of amino acids to form polypeptides.* It takes place on the ribosomes. The function of the ribosome is to attach to the mRNA, by its small subunit, and to hold the messenger in such a way that its codons can be recognised and paired with the complementary anticodons in the tRNA. A ribosome can accommodate two tRNAs at any one time while their amino acids are being linked together by a peptide bond (Fig. 14.8). As the bond is formed, the ribosome simultaneously moves one triplet further along the messenger. The tRNA on the left (Fig. 14.8) is then released, to be used over again, and the next one comes in at the right to pair with the newly-positioned triplet and to add another amino acid to the growing polypeptide chain. As the ribosome moves 'down' the mRNA molecule, amino acids are joined to the growing chain at the rate of 15 per second. The complementary nature of

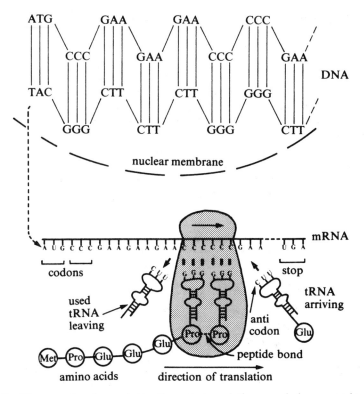

Fig. 14.8 Diagram showing a single ribosome translating a coded message in mRNA into a polypeptide chain. Part of a gene from which the mRNA was transcribed is also shown. The ribosome attaches to the mRNA and then moves along it one triplet at a time. The translation is achieved by the way in which the anti-codons of the tRNAs are paired up with their complementary codons in the mRNA. Two tRNAs are held within the ribosome at any one time. When their amino acids have been linked together by a peptide bond the ribosome moves one triplet to the right. The tRNA on the left is then released and another one moves in from the right to occupy the new exposed triplet (which will be **GAA**). Since each tRNA carries a specific amino acid the transcribed message in the mRNA is thus translated into a particular sequence of amino acids in the polypeptide.

the base pairing between the mRNA codons and the anticodons of the tRNAs ensures that the transcribed message in the mRNA is faithfully translated into the correct sequence of amino acids in the polypeptide product.

Polypeptide chains usually begin with the amino acid **methionine**. This is coded for by the triplet **TAC** in the DNA. It is transcribed as the codon **AUG** in mRNA and recognised by the tRNA with the anticodon **UAC**. The average length of a polypeptide is 333 amino acids. Their sequence is completed when the ribosome comes to one of three special **stop signals—UAA, UGA** or **UAG**. The tRNAs that pair with their codons do not carry an amino acid.

The process by which the transcribed information carried in the base sequence of mRNA is used to produce a sequence of amino acids in a polypeptide chain is known as **translation**.

Once the polypeptide is synthesised it dissociates from the ribosome and is released into the cytoplasm. There it may undergo some **post-translational modification**, such as folding or associating with other polypeptides, to form a functional protein.

Electron micrographs showing the translation of mRNA often reveal long chains of ribosomes (**polyribosomes**). It is thought that these occur because one mRNA strand is being simultaneously translated by a succession of ribosomes. Once the amino acids have been linked by peptide bonds, and their tRNAs released, the beginning of the mRNA strand is free, and can be used by another ribosome. As this second one moves along the strand the beginning will become free again and a third ribosome can come into play. A number of polypeptides, all of which are identical, can thus be synthesised from one mRNA molecule at the same time (Fig. 14.9).

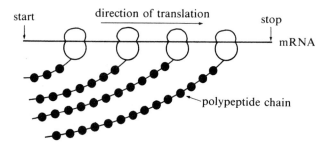

Fig. 14.9 Diagram showing how several ribosomes (a polysome) can simultaneously translate the same mRNA strand. When the first ribosome has translated a number of codons near the start of the messenger, and moved along the strand, a second ribosome can attach itself and move along behind the first one. A third and fourth ribosome, and so on, can attach in sequence as the preceding ones, with their growing polypeptides, move 'down' the strand.

The flow of information

The question we posed earlier (p. 193) about how the genetic information in DNA may be used to determine the structure of proteins has now been answered. The answer may be summarised in terms of the **flow of information**:

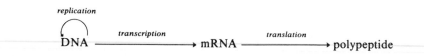

An important point to note here is that the flow of information takes place at all stages by the principle of complementary pairing of nucleotide bases.

Prokaryotes and eukaryotes

The flow of information takes place in the same general way in both prokaryotes and eukaryotes, but there are certain differences between the two groups: these are described in Box 14.2.

Box 14.2   Flow of information in prokaryotes and eukaryotes

In the prokaryotes there is no nuclear membrane and therefore no need for the transcribed mRNA to be moved from the nucleus to the cytoplasm—it is already in the cytoplasm. In bacteria, and the other prokaryotes, translation and transcription are *coupled*. As the mRNAs are being transcribed from the DNA, the ribosomes attach to them and translation begins even before transcription is complete. In the eukaryotes, as we have described, this is not the case: transcription is completed first and then the mRNA moves out of the nucleus where it is translated. Furthermore it is now known that in eukaryotes the mRNA transcripts are *modified* by the addition of **caps** and **tails**, before they leave the nucleus. Before transcription is completed a cap, consisting of a modified guanine, is attached to the 5′ end of the transcript by a bond that includes three phosphate groups. The guanine is modified by **methylation**, i.e. the addition of extra methyl (CH_3) groups. The function of the cap is not fully understood, but it appears that it may act as a signal to promote translation.

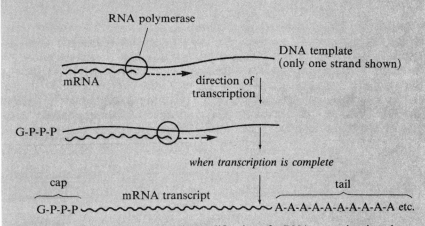

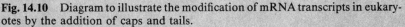

Fig. 14.10   Diagram to illustrate the modification of mRNA transcripts in eukaryotes by the addition of caps and tails.

Most mRNA transcripts in eukaryotes are also modified at the 3′ end. Almost immediately after translation is completed a tail, of a hundred or so adenine nucleotides, is added. The adenines are collectively called poly-A. The poly-A tail is thought to act as a signal for export of the messenger out of the nucleus, and also to protect it from breakdown by enzymes in the cytoplasm. Messengers without a tail last only minutes in the cytoplasm, whilst those which have a tail may last for several days. The addition of caps and tails is illustrated in Fig. 14.10. (See also Box 14.3 for further details on processing of mRNA in eukaryotes.)

Deciphering the code

The genetic code has been completely deciphered and all the 64 codons have been assigned to their respective amino acids or stop signals. This was done in the 1960s in a series of experiments devised by Nirenberg and Mathaei.

They constructed synthetic mRNAs of simple known base sequence, and then introduced these mRNAs into a cell-free extract of *E. coli* in a test tube. The cell-free extracts contained the necessary 'machinery' of protein synthesis, i.e. ribosomes, tRNAs, enzymes and amino acids. Polypeptides were synthesised and isolated using this cell-free system. When a synthetic messenger comprising only uracil nucleotides (poly-U) was used, a polypeptide made entirely of phenylalanine residues was the result (Fig. 14.11).

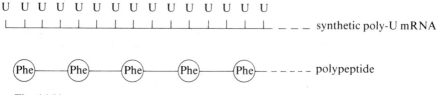

Fig. 14.11

The code for phenylalanine was therefore shown to be **UUU**. Similar tests, with poly-A and poly-C, revealed that **AAA** coded for lysine and **CCC** for proline. Poly-G didn't work: it folded up in a way that would not allow for translation. Synthetic messengers with mixtures of the nucleotides in various ratios eventually revealed the assignments of all the codons, as given in Fig. 14.12. In referring to the genetic code we are therefore talking about the codons in mRNA: these are complementary to the sequences in the transcribed coding strand of DNA and identical with those in the non-transcribed strand.

Second base

		U	C	A	G	
First base	U	UUU ⎱ Phe UUC ⎰ UUA ⎱ Leu UUG ⎰	UCU ⎱ UCC ⎰ Ser UCA ⎰ UCG ⎰	UAU ⎱ Tyr UAC ⎰ UAA Stop UAG Stop	UGU ⎱ Cys UGC ⎰ UGA Stop UGG Trp	U C A G
	C	CUU ⎱ CUC ⎰ Leu CUA ⎰ CUG ⎰	CCU ⎱ CCC ⎰ Pro CCA ⎰ CCG ⎰	CAU ⎱ His CAC ⎰ CAA ⎱ Gln CAG ⎰	CGU ⎱ CGC ⎰ Arg CGA ⎰ CGG ⎰	U C A G
	A	AUU ⎱ AUC ⎰ Ile AUA ⎰ AUG Met	ACU ⎱ ACC ⎰ Thr ACA ⎰ ACG ⎰	AAU ⎱ Asn AAC ⎰ AAA ⎱ Lys AAG ⎰	AGU ⎱ Ser AGC ⎰ AGA ⎱ Arg AGG ⎰	U C A G
	G	GUU ⎱ GUC ⎰ Val GUA ⎰ GUG ⎰	GCU ⎱ GCC ⎰ Ala GCA ⎰ GCG ⎰	GAU ⎱ Asp GAC ⎰ GAA ⎱ Glu GAG ⎰	GGU ⎱ GGC ⎰ Gly GGA ⎰ GGG ⎰	U C A G

Third base

Fig. 14.12   The genetic code. The code is given here, as is usual, in terms of the mRNA codons.

Features of the code

It is not necessary of course to remember the assignments of the various codons, but there are certain features of the genetic code which should be noted:

1. Certain amino acids (methionine and tryptophan) are coded for by only one codon, but most are coded for by several (Fig. 14.12). The code is therefore said to be **degenerate**, because it contains more information than is required, i.e. there are more codons than there are amino acids. This degeneracy is mostly accounted for by variations in the third base of the codon: for example **GCU, GCC, GCA** and **GCG** all code for alanine. The family of tRNAs that carry alanine therefore recognise only the first two bases in the mRNA codon, and the third one is relatively unimportant.
2. Some triplets do not code for any amino acids: these are the **stop codons**, or **chain-terminating codons, UAA, UAG** and **UGA**. They play a vital role as signals to terminate translation and to bring an end to the formation of a polypeptide chain. The stop codons can be thought of as a form of punctuation which separates one gene from its neighbours.
3. The code is continuous and **non-overlapping**. If it were overlapping, so that the beginning of one triplet was also used for the end of the one preceding it, then a mutation (say a base pair substitution) in one codon could affect more than just one amino acid in a polypeptide—and this is not the case.
4. The code is **universal** and the same codons are used to specify the same amino acids in all living organisms—*with one exception*. In the mitochondrial DNA of man, and of yeast, certain chain-terminating codons are read as amino acid-specifying codons, and vice versa.

What is a gene?

Thus far we have only been able to envisage the gene as a particulate structure which we have referred to as a 'unit of heredity'. This definition was based on our observation of the way in which characters are inherited in crossing experiments. Patterns of inheritance could only be explained if the characters concerned were determined by particular unit structures, genes, which were present as allelic pairs at corresponding loci within homologous chromosomes. The substance which makes up the chromosomes and which carries the information in the genes, is DNA. Now that we know about the structure of DNA, and about the way it acts to determine characters, we can say what the gene is in precise biochemical terms:

A **gene** is *a sequence of nucleotide pairs along a DNA molecule which codes for an RNA or polypeptide product.*

Both strands of the double helix make up the gene. Only one strand, the **transcribed coding strand**, contains the information which is directly utilised for RNA or amino acid assembly. The other strand, i.e. the **non-transcribed** one, contains nucleotide sequences which are complementary to those of the coding strand and is used for replication. The transcribed coding strand is not necessarily the *same* strand in all of the genes along a chromosome: it switches from one to the other. The structure of the gene is shown diagramatically in Fig. 14.13.

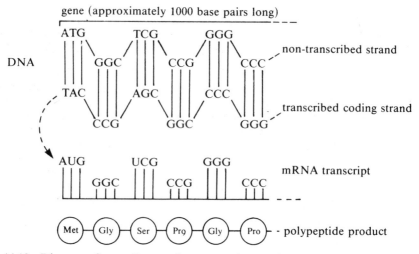

Fig. 14.13 Diagram of a small part of a structural gene showing the organisation of the base triplets and their relationship to the mRNA transcript and the polypeptide product.

Genes which code for polypeptides fall into two main classes. These are (1) **structural genes**, which code for functional proteins (enzymes, hormones, components of cell structure, antibodies, storage proteins, etc.), and (2) **regulatory genes**, which serve to control the activity of other genes—we deal with these in Chapter 17. Polypeptide chains vary in their length and so too do the genes which code for them. The average size of a polypeptide chain is 333 amino acids, so the average size of a structural gene is of the order of 1000 nucleotide pairs of DNA.

Transfer and ribosomal RNA molecules are also coded for by genes, but they are not synthesised in the same way that proteins are. The tRNA and rRNAs are made directly by transcription from the DNA, in the same way as mRNA. Transfer RNA molecules are about 80 nucleotides long, and the genes which code for them are therefore of corresponding length in terms of DNA. Ribosomal RNAs are of three sizes (p. 196)—100 bases, 1500 bases and 3000 bases, and their genes consist of corresponding lengths of DNA nucleotides. The genes which code for rRNAs are present in multiple copies, and in eukaryotes they are localised at a special region in the chromosome called the **secondary constriction**, or **nucleolus organiser**.

Split genes and redundant DNA

It is impossible to give a complete account of the structure and action of DNA, and of genes, because new facts and surprises are emerging all the time as research into molecular-genetics continues to race ahead. New techniques of genetic engineering (Chapter 22) developed in the 1970s made it possible to study the structure and organisation of genes, and of chromosomes, in great detail. Individual genes, and large stretches of chromosomal DNA, can now be sequenced, and their nucleotide composition analysed and compared with their RNA transcripts and protein products (Chapter 22). It turns out that most eukaryotic genes are 'split', and have far more DNA within them than is actually required to code for the amino acids in their protein products, i.e. they contain stretches of **non-coding DNA**. It also transpires that large regions of the chromosomes do not appear to contain any genes at all: they are composed of stretches of repetitive DNA, i.e. small sequences of bases which are present as millions upon millions of tandemly repeated copies. This repetitive (or **redundant**) DNA has no known function. Some of it is known to be transcribed, in certain organisms, but we have no idea what the protein products are (if any), or what purpose they serve within the cell. Further details are given in Box 14.3.

Box 14.3 Split genes and redundant DNA

In eukaryotes, as we have explained, transcription and translation are separate events. Transcription takes place within the nucleus and the mRNA then moves out into the cytoplasm for its translation. It turns out that there is also another event—a **processing stage**—which happens while the mRNA transcript is still within the nucleus. This is necessary because many eukaryote genes are 'split' and contain several stretches of additional non-coding DNA, called **introns**, over and above that required to specify the amino acids of their proteins. The coding sequences are known as **exons**. When transcription takes place all of the DNA bases are copied in the mRNA transcript. The processing stage then occurs and snips out the introns, by a means of a splicing enzyme, to form the 'mature' mRNA which moves to the cytoplasm for translation. Only the coding triplets in the exons are used to determine the amino acid sequence of the polypeptide (Fig. 14.14). The function of the DNA in the non-coding introns is not known. Splicing of the primary transcript takes place after the addition of the caps and tails that we mentioned earlier (p. 199).

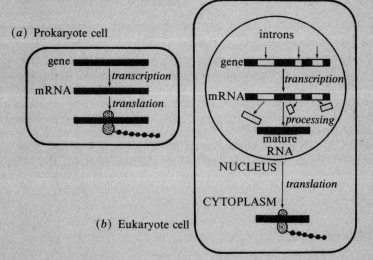

(a) Prokaryote cell

gene

mRNA

transcription

translation

introns

gene

mRNA

transcription

processing

mature RNA

NUCLEUS

translation

CYTOPLASM

(b) Eukaryote cell

Fig. 14.14 The organisation of genes is different in (a) prokaryotes and (b) eukaryotes. Eukaryotes have large amounts of non-coding DNA located in several introns within their genes. The non-coding DNA is transcribed but the transcribed sequences are then processed out of the mRNA, by a splicing enzyme, before translation takes place.

Introns form only one category of the redundant DNA which is a general feature of eukaryote genomes. Other kinds of non-coding DNA include spacer sequences which occur between certain genes, and short repetitive sequences of bases, too small to be genes, which are present as millions of repeated copies scattered throughout the chromosomes. No one knows why the nucleus contains so much apparently useless DNA. It is one of the major unsolved paradoxes of genetics.

Genes and characters

In this chapter we have now seen how a simple biochemical defect, such as an enzyme deficiency, can lead to a visible change in the phenotype of a character. This is evident in the metabolic disorders in man and the auxotrophic mutants in *Neurospora*. Since the enzymes concerned are directly controlled by individual genes, the relation between the gene and the character is obvious. We will have more to say on this matter in Chapter 15, after we have dealt with the subject of mutation.

Summary

Studies on metabolic disorders in man, and the biochemical genetics of *Neurospora*, led to the idea that genes act by determining the structure of enzymes, and thereby control all the biochemical activities of the cell. It was subsequently shown that genes determine the structure of all proteins, including enzymes, and that one gene is responsible for specifying the sequence of amino acids of one polypeptide chain. The information in the DNA is encoded in such a way that one triplet of bases carries the information specifying one amino acid. When the information is used it is first of all transcribed into a molecule of single-stranded messenger RNA, and then translated into protein through the involvement of ribosomal and transfer RNA. The process is known as protein synthesis. The ribosomal and transfer RNAs are themselves transcribed directly from DNA. The gene is thus defined as a sequence of nucleotides of DNA that codes for an RNA or protein product. In eukaryotes there is a large amount of non-coding redundant DNA, the function of which is unknown.

Further reading

Beadle, G. W. (1948), The genes of men and moulds, *Scientific American* **179**, No. 9, 30–40.
Chambon, P. (1981), Split genes, *Scientific American* **244**, No. 5, 60–71.
Crick, F. H. C. (1962), The genetic code, *Scientific American* **207**, No. 4, 66–74.
Crick, F. H. C. (1966), The genetic code: III, *Scientific American* **215**, No. 4, 55–62.
Nirenberg, M. W. (1963), The genetic code: II, *Scientific American* **208**, No. 3, 80–94.

Questions on Chapters 13 and 14

1 The diagram below represents the basic chemical unit from which the nucleic acids DNA and RNA are formed.

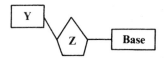

(a) What is the name of:
    (i) the chemical unit illustrated;
    (ii) the component labelled **Y**;
    (iii) the component labelled **Z**?
(b) Which of the bases found in DNA does not occur in RNA?
(c) If adenine is one of the bases on the code chain of a molecule of DNA what will be:
    (i) the base paired with it in the DNA;
    (ii) the base in messenger ribonucleic acid which is paired with it;
(d) the base in transfer ribonucleic acid which pairs with the answer given in (b)?
                                  (O.L.E. Biol., 1981)

2 (a) In the DNA molecule there are four major bases, two purines and two pyrimidines.
These are

       purine   pyrimidine
    (i) ....   (i) ....
    (ii) ....   (ii) ....
(b) Name the sugar in
    (i) DNA............   (ii) RNA............
(c) What is a 'nucleotide'?
(d) What are the possible base pairings of the four bases?
(e) One of the bases in DNA is not represented in RNA. Name the base not represented and the base that replaces it.
    (i) Base not represented...........   (ii) Base that replaces it...........
(f) Name three types of RNA involved in protein synthesis.   *(L. Biol., 1981)*

3 (a) Describe the structure of (i) DNA and (ii) chromosomes.
(b) Give a brief account of the way in which the information contained in DNA is converted into new protein molecules.   *(C. Zool., 1982)*

4 Compare the structure of DNA with that of protein and explain the biological significance of their similarities and differences.   *(J.M.B. Nuffield Biol., 1982)*

5 What part is played by nucleic acids in the synthesis of proteins and by proteins in the synthesis of nucleic acids?   *(O.L.E. Zool., 1981)*

6 To what extent has the genetic code been cracked?   *(O.L.E. Bot. S., 1979)*

7 Write an essay entitled "Ribosomes and their function". *(W.J.E.C. Biol. S., 1982)*

8 (a) The diagram below shows the sequence of nitrogenous bases on part of a strand of DNA and the corresponding region of a strand of messenger RNA. Using the letters provided in the key, indicate on the diagram the complementary bases which would be found on the strand of messenger RNA.

Strand of DNA

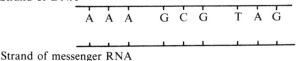

Strand of messenger RNA

KEY: **A** = adenine; **G** = guanine; **T** = thymine; **U** = uracil; **C** = cytosine.

(*b*) State briefly why it is thought that the arrangement of bases of nucleic acids (DNA or RNA) is in the form of a 'triplet code'.

(*c*) (i) What is an alteration in the sequence of bases called?
(ii) Give *one* example of how this alteration might be caused.

(*d*) Name *two* groups of organisms in which the DNA is not enclosed within a distinct nuclear membrane.

first group ....
second group ....   (*L. Biol., 1983, part question*)

9 Nucleic acids are polymers of nucleotides. Each nucleotide consists of a purine or pyrimidine base and a sugar esterified with a molecule of phosphoric acid.

(*a*) Name the two kinds of sugar that occur in nucleotides.

(*b*) How many atoms of carbon are contained in each of these sugars?

(*c*) Name:
(i) the three bases common to DNA and RNA;
(ii) the base that is specific to DNA;
(iii) the base that is specific to RNA.

(*d*) In the double helix of DNA two parallel strands are linked together by hydrogen bonds between adjacent bases. Name the Nobel Prize winning scientists who deduced this model structure for DNA.

(*e*) Give the bases which pair together in the double helix
(i) ....
(ii) ....

(*f*) In the genetic code, what is the minimum number of bases necessary to code for an individual amino acid?

(*g*) What is the name given to the type of RNA that is formed along one strand of the DNA helix and carries the coded information for protein synthesis?

(*h*) By means of a diagram show how this type of RNA copies the base pattern of the DNA strand.

(*i*) What is the name of the type of RNA that combines with individual amino acids and organises the synthesis of the peptide chain?

(*j*) In which organelle does the majority of protein synthesis occur?

(*k*) What is the name of the type of RNA that forms the structural framework of this organelle?   (*W.J.E.C. Biol., 1978*)

15
Gene mutation

In the preceding chapters we have seen how the structure of DNA allows for its self-replication and for the encoding and utilisation of genetic information during gene action. The processes of replication, transcription and translation take place with an extraordinarily high level of fidelity. This is due to the very specific way in which the nucleotide bases bond together as complementary pairs. The stability of DNA, the accuracy with which it is replicated and the way the information encoded within it is used, are the essence of heredity. This constancy enables species to maintain their genetic programs and to preserve their identity over countless cycles of development and reproduction.

But DNA replication is not perfect. If it were, diploid organisms would have two identical copies of each gene at all their different loci, and every individual of a species would be of the same genotype. There would be no genetic variation and no capacity to evolve and to adapt to long term changes in the environment. By the same token there would be no character differences with which the geneticist could work and no alleles with which to study the inheritance, the structure and the action of the genetic material.

The capacity for change, and for heritable variation, comes about by *mutation.*

Gene mutation

A **mutation** is *a sudden heritable change in the genetic material.* **Gene mutations** (or point mutations) result from changes occurring within single genes, and they give rise to new alleles. In referring to 'change', we mean some departure from the normal wild-type form of a gene, or from the form normally present in an experimental laboratory strain of an organism which we use as our standard type.

How do gene mutations arise?

Gene mutations can arise spontaneously or they may be induced.

Spontaneous mutations result from errors in the replication of DNA. We have only mentioned one of the enzymes involved in this process, DNA-polymerase (p. 179), but in reality it requires a vast number of different enzymes. These enzymes make very rare mistakes and this leads to changes in the base composition of DNA.

Induced mutations are changes in DNA which are caused by the effects of 'mutagens'. We will deal with the action of mutagens later (p. 217). Changes in the base composition of DNA, spontaneous or induced, constitute the molecular basis of gene mutation.

Molecular basis of gene mutation

There are three main ways in which the base composition of the DNA of a gene may be altered by mutation—these are shown in Fig. 15.1. The changes may involve just one, or more than one, nucleotide pair. These changes can be detected because they affect the information carried in the genetic code. If we recall the pathway of information flow given in the previous chapter (p. 198) it is obvious that a change in the nucleotide sequence of a gene can result in a corresponding change in the amino acid sequence of its polypeptide product. Looking at the changes in terms of gene action and protein products we can classify mutations into several categories.

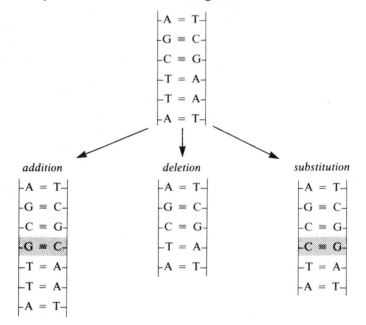

Fig. 15.1 Alterations in the base composition of DNA causing gene mutation.

Samesense mutations

A **samesense mutation** is *a base substitution in a DNA triplet which does not alter the amino acid sequence of a polypeptide.*

These mutations can happen because the genetic code is degenerate. The majority of amino acids are coded for by several mRNA codons which vary in their third base (p. 200). Proline, for instance, is coded for by **CCU, CCC, CCA** and **CCG**. If a mutation in the third base of a DNA triplet changes an mRNA codon from **CCA** to **CCC** it will not make any difference to the polypeptide which is formed.

Missense mutations

A **missense mutation** is *a base-pair substitution in a gene which results in one amino acid being changed for another one at a particular place in a polypeptide.*

For example, if a mutation in DNA alters an mRNA codon from UUU to UGU then at a corresponding place in the polypeptide product phenylalanine will be replaced by cysteine. Such changes may have relatively little effect upon the function of the protein concerned, or they may be quite serious as in the case of sickle-cell anaemia (p. 193).

Nonsense mutations

A **nonsense mutation** is *any mutation* (substitution, addition or deletion) *that changes an amino acid-specifying codon into a chain-terminating codon* (**UAG, UAA** or **UGA** in mRNA).

This results in the premature termination of a polypeptide chain. The way in which a substitution can cause a nonsense mutation is illustrated in Fig. 15.2.

Normal code in mRNA (*direction of reading* →)

AUG UAU CUG AUU UCA AAA CCC   ...   etc. (say 300 codons)

Met   Tyr   Leu   Ile   Ser   Lys   Pro   —   etc. (300 amino acids)

Substitution   ↓

AUG UAU CUG AUU UGA AAA CCC   ...   etc.

Met   Tyr   Leu   Ile   STOP

Fig. 15.2

Frameshift mutations

A **frameshift mutation** is *a mutation resulting from the addition or deletion of one or more nucleotides, other than in multiples of three, that causes the gene to be 'misread'.*

The **reading frame** describes the way in which the code in mRNA is read as triplets by the ribosome (Fig. 14.8). The insertion or deletion of bases, other than in multiples of three, results in a shift in the reading frame and creates an entirely new sequence of codons—some of which are nonsense codons. The gene product is therefore non-functional because it has the wrong sequence of amino acids in the polypeptide and the chains are also terminated prematurely (Fig. 15.3*a*, overleaf). The insertion or deletion of three bases does not affect the reading frame, but it does result in the inclusion or loss of one amino acid from the polypeptide (Fig. 15.3*b*). It was Francis Crick who first discovered the effect of adding or deleting bases in threes and this gave him the vital evidence that the code was organised as *triplets* of bases.

In the above account we have concentrated on genes which code for polypeptides. It should also be mentioned that mutations occur in genes coding for rRNAs and tRNAs as well.

Somatic cells and gametes

It is important before proceeding any further to distinguish between mutations which arise in somatic cells and those which occur in the germ line or in the gametes themselves—because the consequences are quite different. Somatic

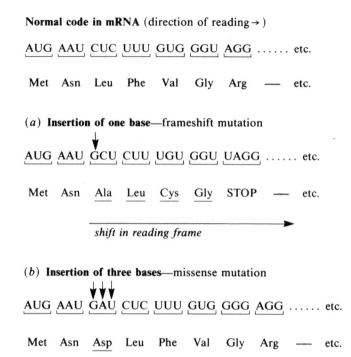

Normal code in mRNA (direction of reading →)

AUG AAU CUC UUU GUG GGU AGG ...... etc.

Met Asn Leu Phe Val Gly Arg — etc.

(*a*) **Insertion of one base**—frameshift mutation

AUG AAU GCU CUU UGU GGU UAGG ...... etc.

Met Asn Ala Leu Cys Gly STOP — etc.

shift in reading frame →

(*b*) **Insertion of three bases**—missense mutation

AUG AAU GAU CUC UUU GUG GGG AGG ...... etc.

Met Asn Asp Leu Phe Val Gly Arg — etc.

Fig. 15.3 (*a*) Frameshift mutations result from the insertion or deletion of a number of nucleotides, other than multiples of three. The shift in the reading frame leads to the generation of an entirely different set of codons, including nonsense codons. In most instances, therefore, the polypeptide product is completely non-functional.
(*b*) When three bases are added, or deleted, only one amino acid is affected and the reading frame remains in phase.

cell mutations affect only a few of the cells in an individual higher organism, and the majority of the tissues are unaffected and are of normal phenotype. Such mutations may be quite trivial in their effects, and in a sexually-reproducing species they are not inherited. Some of them can be serious to the individuals concerned if they result, for instance, in a detrimental condition such as a malignant tumour in an animal. Mutations in the germ line, or in the gametes, are of much greater significance since they will be present in all of the cells of the progeny that arise from the affected gametes. They will be transmitted to future generations and may make a contribution to the genetic variability of the species (Chapter 20).

Somatic cell mutations can be of importance in asexual reproduction if the cells or tissues involved happen to be in that part of the organism which becomes detached to give a new individual, e.g. asexual spores in fungi (see Chapter 18).

Gene mutations and character differences

Character differences result from gene mutations. In this section we will first of all consider how a mutation affects a character. We will then explain why it is that most mutations are deleterious and recessive. Finally we will itemise the kinds of effects that they have upon the phenotype.

How do mutations cause character differences?

In the present context we can consider that the expression of genetic information involves two steps. The first one is the transcription and translation of the information into a polypeptide. The second step is concerned with the action of the polypeptide:

$$\text{gene} \xrightarrow{\textit{information flow}} \text{polypeptide} \xrightarrow{\textit{action}} \text{character}$$

When a mutation takes place the information in the gene is changed, and it is then expressed differently:

$$\text{mutation} \rightarrow \begin{matrix} \text{new} \\ \text{allele} \end{matrix} \xrightarrow[\textit{information}]{\textit{different}} \begin{matrix} \text{altered} \\ \text{polypeptide} \end{matrix} \xrightarrow[\textit{action}]{\textit{different}} \begin{matrix} \text{character} \\ \text{difference} \end{matrix}$$

In a diploid there are two alleles being expressed in the same cell. When they are identical (**AA** or **aa**) we have no difficulty in understanding how they contribute to the character. When they are different (**Aa**) we have to explain how they act, or interact, together. The details of the explanation are given in Box 15.1. We also have to remember at this stage that the actual phenotype we observe depends upon the way in which the genetic effects interact with the environment.

We have already discussed the mutation which causes sickle-cell anaemia (Chapter 14, p. 192). In this case we know exactly what biochemical changes are involved in the mutant allele (single base pair substitution), how it alters the β polypeptide product (single amino acid substitution) and what the consequences are to the phenotype. This is one of the few mutations that we understand so thoroughly.

Box 15.1   Mutations and character differences

The simplest situation with which to explain how mutations affect characters is that in which the change results in complete loss of enzyme activity. This could arise either because the polypeptide is not formed, or because it is altered in such a way that it cannot function normally. The change can be symbolised as **A → a**. A is the normal form of the gene coding for an enzyme which converts a precursor substance (call it X) into the pigment of a flower—say a purple-coloured one—and **a** is the recessive allele which is unable to give rise to a functional enzyme. In the dominant homozygote (**AA**), both alleles will produce the enzyme giving a normal purple-flowered phenotype. The recessive homozygote (**aa**) will not be able to metabolise the precursor and the flower will exhibit a white mutant phenotype. The heterozygote (**Aa**) shows full dominance because the normal allele produces sufficient enzyme to make all of the pigment required for purple flowers (Fig. 15.4).

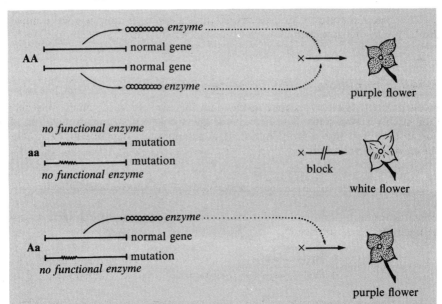

Fig. 15.4 Explanation of how a mutation can cause a character difference, and what is meant by *dominant* and *recessive* in molecular terms. **A** is the normal dominant allele and **a** is the recessive mutant allele.

Another situation is that involving incomplete dominance. Suppose we are again dealing with a gene controlling a flower colour phenotype, such that **RR** gives red flowers, **Rr** gives pink and **rr** is white. To explain this effect we simply have to postulate that the normal allele (**R**) in the heterozygote produces insufficient enzyme to synthesise the full complement of red pigment.

The co-dominance that we find in the ABO blood group system in man (Chapter 10) is a kind of protein polymorphism which is not evident in the visible phenotype. This character is explained because the heterozygote ($I^A I^B$) has two variants of a normal gene both of which produce a product:

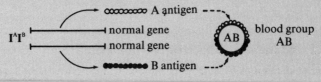

Fig. 15.5

We have confined ourselves here to a few simple examples. There are many other complications that could be considered, including those where we have interactions between two genes controlling the same character (Chapter 9). We must also remember that mutations do not necessarily have all-or-nothing effects. Their consequences to the phenotype depend upon the kind of change that has taken place within the gene and the way in which the amino acids are altered in the gene product.

Missense mutations are generally less serious than nonsense and frameshift mutations for the reasons given earlier in the chapter.

Why are mutations mainly recessive and deleterious?

Each gene that codes for a polypeptide is made up of a certain sequence of nucleotide pairs of DNA. The code is organised in triplets of bases which are transcribed and translated into a corresponding sequence of amino acids in the protein product. The correct sequence of amino acids is essential for the protein to function properly (although *some* minor changes can be tolerated).

Since a mutation is a random change in the structure of a gene it follows that it will result in a correspondingly random change in the structure of the protein which it determines. As a result the protein will suffer some impairment to its function (missense mutation), or else it will not be produced at all (nonsense mutation). By analogy, if we were to cause a random change in the circuitry of a modern television receiver by poking about in the back with a screwdriver, we would most likely damage the quality of the picture, or else lose it completely. It is most unlikely that we would bring about any improvement! Mutations are therefore mainly deleterious because they upset the normal working of a gene. They are usually recessive for the simple reason that in a heterozygote a normal allele of the gene is present as well. The mutant one is hidden by virtue of the fact that it does nothing, or determines a non-functional protein (Box 15.1).

Forward and back mutations

The change from a normal to a mutant form of a gene is known as a forward mutation ($A \rightarrow a$). A change in the opposite direction, from the mutant back to the normal form, is a back (or reverse) mutation ($a \rightarrow A$). The forward mutation rate of a gene is much higher than the back mutation rate. The reason for this is that a forward mutation (say a substitution) involves a random change anywhere within the 1000 base pairs of the gene. To reverse this first mistake, the second random change must involve the *same single base pair* (or another which compensates for the change), rather than *any* of the pairs, and the chances of this happening are clearly much lower.

Types of mutations

Mutations are classified into a number of different types. This is purely for convenience so that we can discuss them and write about them in a way that relates to the kinds of effects they have upon the phenotype.

Morphological mutations   These are the most obvious types and they affect the form (='morph') of the organism. Included in this category are all of Mendel's pea characters and many of those which we have encountered in *Drosophila* (white eyes, vestigial wings, curly wings, ebony body, etc.). We also include here characters to do with appearance in man (eye colour, albinism), comb form in fowl and any other characters which affect the size, shape or colour of an organism.

Lethal mutations   Lethal mutations kill an organism. An albino plant which is lacking in chlorophyll is a good example. In many cases though, we have no idea why the mutation causes death—we simply classify it by its effect.

Biochemical mutations   We have dealt with several of these in bacteria, fungi and man. They are identified by the loss, or defect, in some specific biochemical function of a cell.

Conditional mutations   These are mutations which are expressed only under certain conditions, e.g. coat colour pattern in the himalayan rabbit (Fig. 10.10).

Regulatory mutations   Mutations in genes which control the activities of other genes (Chapter 17).

Protein polymorphisms   There are some mutational changes that are so subtle as to have no obvious visible effect upon the phenotype at all. They can only be detected at the level of variation in the protein products of gene action. Their detection depends on the technique of **protein electrophoresis** to separate out the various forms on the basis of small differences in their electrical charges. The differences result from minor changes (missense mutations) in amino acid composition which have little if any effect upon the function of the proteins. In these cases it is often impossible to decide which is the normal form of the gene and which the mutant—especially if there are two alleles which are both equally common in a population.

Detection of mutations

Mutations in haploid organisms are easily detected, because there are no complications due to dominance and they are expressed immediately in the phenotype. In diploids a dominant mutation which arises in a gamete will also show up straight away in the progeny in which it is inherited. A recessive, on the other hand, will not be apparent in the first generation (unless it is sex-linked) (Fig. 15.6).

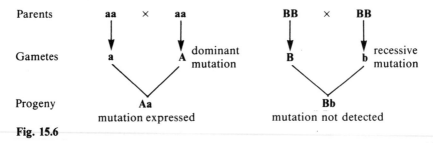

Fig. 15.6

The recessive mutation may go undetected for several generations, even if it is lethal, until two heterozygotes mate together. It will then segregate out as a homozygote (**bb**).

In higher organisms, it is often difficult to find mutations unless they are very distinctive in appearance and easily spotted by eye—for the simple reason that they are rare (see below). Microbes, on the other hand, are much easier to deal with, as we have already explained in Chapter 12. To detect mutations in bacteria which give rise to resistance to antibiotics, for instance, it is simply necessary to grow up several million cells on an agar medium containing the antibiotic (e.g. penicillin). All the normal cells are killed and any colonies which survive are derived from mutant individuals with penicillin resistance. Where a *selective* system of this kind can be used, the odd mutant can be screened out from amongst millions of normal cells.

At one time it was suggested that such screening procedures were not actually detecting spontaneous mutations, but that the antibiotic substances themselves were causing the cells to mutate. This idea was disproved by Joshua and Esther Lederberg who devised a simple experiment to show that mutations giving resistance to certain antibiotics were present in the bacterial cultures whether they were grown on the antibiotics or not. In other words, they confirmed that **mutations are random changes** in the genetic material: the details are given in Box 15.2.

Box 15.2 A replica-plating test to show that mutation is a random process

In 1952 Joshua and Esther Lederberg devised a simple test to show that mutations arise in bacteria by random processes and are not induced in response to a selective agent such as an antibiotic. They prepared a master plate containing several million colonies of *E. coli* on a complete agar medium. This medium contained no antibiotics. A velvet pad was then used to press onto the surface of the culture and to transfer a replica of the colony formation onto fresh plates of selective agar containing the antibiotic. A few colonies with antibiotic resistance grew on the new plates. Because of the way in which the colonies had been replica-plated it was possible to see that the *same* colonies were growing on each of the freshly inoculated plates. In other words the mutations must have arisen spontaneously as random changes on the original master plate, and not in response to the antibiotics—otherwise they would not have been in identical places on the replicas. The experiment is illustrated in Fig. 15.7.

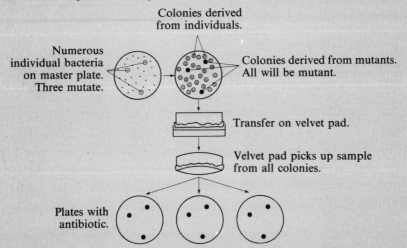

Colonies derived
from individuals.

Numerous
individual bacteria
on master plate.
Three mutate.

Colonies derived from mutants.
All will be mutant.

Transfer on velvet pad.

Velvet pad picks up sample
from all colonies.

Plates with
antibiotic.

Fig. 15.7 Some cells from each colony will get transferred to each of the three replica plates. Only those resistant to the antibiotic will grow. Because the resistant colonies are in identical positions it shows that they must have originated as mutations in single cells in the original master plate—and not in response to the antibiotic.

Mutation rates

How common are gene mutations?

Mutation rate is defined as *the number of mutations in a given gene per unit of time*, where time is measured as cell divisions. Mutation rates are difficult to measure in most organisms (except for viruses, bacteria and cell cultures) because we cannot determine the number of divisions that have given rise to a population of cells. In practice therefore, and especially in higher organisms, we use **mutation frequency** as an estimate of **mutation rate**.

Mutation frequency is simply *the frequency with which a mutation is found in a sample of cells or individuals.* It is normally expressed as the number of mutations per million gametes, and if we make the appropriate testcrosses this is not too difficult to estimate (Table 15.1).

Table 15.1   Estimates of spontaneous mutation frequencies for genes in various organisms.

Organism	Gene or character	Mutations per 1 000 000 gametes
Diplococcus pneumoniae	penicillin resistance	0.1†
E. coli	resistance to bacteriophage T1	0.03†
Drosophila	$e^+ \rightarrow e$ (ebony body)	20
	$ey^+ \rightarrow ey$ (eyeless)	60
	$w^+ \rightarrow w$ (white eye)	40
Mouse	$C \rightarrow c$ (albino)	10
Maize	$R \rightarrow r$	492
	$Pr \rightarrow pr$ (purple)	11
	$Su \rightarrow su$ (sugary)	2.4
	$Sh \rightarrow sh$ (shrunken)	1.2
Man	muscular dystrophy	4–10*
	haemophilia	2–4*

* Range of values based on several estimates from pedigree records.
† Based on cell counts.

As far as we know, all genes mutate. The rate at which it happens spontaneously, under natural conditions, seems to vary quite widely between different organisms and different genes. In higher organisms the frequency ranges between one in a million gametes (i.e. 1×10^{-6}) to 6×10^{-5}, leaving aside the exceptionally high value of about 5×10^{-4} for the **R** locus in maize. In bacteria the frequencies are lower (1×10^{-7} to 3×10^{-8}) (Table 15.1).

A number of genetic and environmental factors affect mutation rates.

Factors affecting mutation rates

Different genes mutate at different rates

A given gene has a fixed mutation rate under standard conditions, but these rates vary from one gene to another. One reason for this could be the difference in sizes of different genes—the larger the gene the greater the chance of a mutation.

Background genotype

A given gene may have different rates of mutation in different genetic backgrounds. The colour gene mutation **R**→**r** in maize happens three times more frequently in the Cornel strain than it does in the Columbia strain.

Transposable elements

Some genetic elements can cause other genes to mutate at an enhanced rate. One such case is that of the controlling elements discovered in maize by Barbara McClintock. These are short sequences of DNA bases which can move about (transpose) from one part of the chromosome to another! They are usually called **transposable elements** (or **transposons**) and when they insert themselves into a gene they upset its structure and cause a mutation. Transposable elements are also well known in *Drosophila* and *E. coli*. Their existence was not believed for a long time because they don't fit in very well with the chromosome theory of heredity.

Temperature

Spontaneous mutation rates can sometimes be altered by keeping organisms at higher or lower temperatures than normal.

Mutagens

A number of environmental agents, known as **mutagens** (radiation and chemicals), cause genes to mutate at rates which are much higher than their spontaneous level. Mutations which arise in this way are known as **induced mutations**. Induced mutations are important for several reasons: they provide a useful tool for the geneticist; they may be useful to man in relation to plant and animal breeding (Chapter 21); and they constitute an environmental hazard because they damage the genetic material of all living organisms.

In 1927, H. J. Müller found out that X-rays increase the mutation rate in *Drosophila*. He was later awarded the Nobel prize for this important discovery. Independently, and at about the same time, Lewis Stadler demonstrated that X-rays induce mutations in maize. Since that time a large number of different mutagens have become known. We can place them in three categories:

1. Ionising radiations   This category includes X-rays, cosmic rays and radiation coming from various radioactive sources, e.g. alpha (α), beta (β) and gamma (γ) rays. They damage the genetic material directly by breaking it up, or else they break up other molecules (e.g. water) which then become reactive and can then damage the DNA indirectly. With this kind of radiation there is a *linear* relationship between the induced mutation rate and the dose of radiation given (Fig. 15.8). The graph can therefore be extrapolated back through the origin and we can conclude that radiation damage will be occurring even at levels which are too low to be detected experimentally. In other words **there is no such thing as a safe dose**. What is more the effects are *cumulative*, at least to some extent, so that a high dose can be given in the form of several small doses over a period of time.

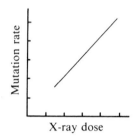

Fig. 15.8 Graph showing the linear relationship between X-ray dose and induced mutation rate.

2. Non-ionising radiations, i.e. ultra-violet light (UV)   UV light is less penetrating than ionising radiations and acts in a different way. It is absorbed by the purine and pyrimidine bases in DNA and modifies them in various ways. One of the main effects is to produce **thymine dimers**. This occurs when two adjacent thymine molecules in one of the DNA strands becomes linked together (T̂ T). The dimer forms a bulge in the helix and this disrupts the bonds between the thymine molecules in the dimer and their complementary adenines (**AA**) in the opposite strand. In the presence of visible light many of these dimers are removed by repair enzymes which make good the damage.

There are two interesting aspects of UV-induced damage that we should mention in relation to the **light repair system**. The first is that when UV light is used to kill bacteria, the cells which have been treated should be kept in the dark. The second aspect concerns a skin cancer in man known as **xeroderma pigmentosum**. This cancer is induced by the UV rays in sunlight and is found in individuals who have a mutation in a gene which determines one of the repair enzymes. The majority of us can tolerate sunbathing because of the highly effective enzyme repair system which makes good the damage we inflict upon ourselves.

3. Chemical mutagens   Chemical **mutagenesis** was discovered by Charlotte Auerbach and J. M. Robson in their experiments with mustard gases during World War II. Their work was classified as secret for several years. Mustard gas is an alkylating agent which affects the base guanine. Alkylating agents can change the pairing specificity of guanine, and so induce a base pair substitution; they can cause it to become unstable so that it is released from the DNA leaving a gap, which may then be filled by another base. There are numerous other chemical mutagens. Many of them are **base analogues**. They have structures very similar to the normal bases and become incorporated into the DNA in place of them during replication. They act through a process known as **tautomerism**, which can cause a change in their pairing specificities. An example is 5-bromouracil (5-BU) which is an analogue of thymine. It has a bromine atom at the ^{5}C position in place of the methyl group (CH_3) in thymine. This bromine atom causes the molecule to change spontaneously from one state to another by the redistribution of its electrons: the change in state is known as a **tautomeric shift**. In the *keto* state BU behaves like thymine and is complementary to adenine during replication (forming two bonds); whereas in its rare *enol* state it forms three bonds and pairs with guanine. This change in pairing specificity can lead to base-pair substitution during DNA replication, and therefore to a mutation (Fig. 15.9).

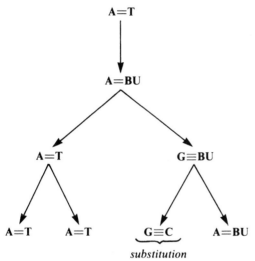

1. normal base pair in DNA

2. bromouracil base analogue (mutagen) incorporated into DNA in place of thymine during replication

3. tautomeric shift in BU changes its pairing specificity from **A** to **G**

4. at the next replication **G** pairs with its complementary base **C** and the substitution is made in one of the daughter molecules

Fig. 15.9

Fate of mutations

The question of what happens to gene mutations once they have arisen in a natural population is dealt with in Chapter 20. Generally speaking most mutations are unfavourable for the reasons given (p. 213), and they are quickly eliminated by natural selection. Recessives in diploids may remain 'hidden' in heterozygotes for several generations before they segregate out as homozygotes. Each population therefore carries a genetic load of deleterious recessive alleles. Mutations which improve fitness will be favoured by selection and will increase in frequency, so that one allele may gradually replace another.

Evolution depends upon the tiny minority of mutations that are not harmful—although what is harmful in one environment may well turn out to be beneficial in another one at a different time. The importance of mutations in the genetic system of variation is discussed further in Chapter 18.

Summary

DNA is a highly stable molecule, but rare errors in its replication produce heritable changes known as gene mutations. These random changes arise spontaneously by the substitution, addition or deletion of nucleotide bases in DNA. The ones which occur in the gametes are more important than those which arise in somatic cells because they are transmitted to future generations. Gene mutations cause a variety of effects upon the phenotype: most are deleterious and recessive. The way in which they affect characters can be simply explained in terms of the biochemistry of gene action. Mutations are much easier to detect in micro-organisms than in higher plants and animals—this is because they are haploid and several millions can be screened using selective procedures to pick out only the mutants. Genes mutate spontaneously at very low rates. The rate can be increased by the use of physical and chemical mutagens. Mutations are important in experimental genetics and as a source of variation in natural populations.

Questions follow Chapter 16.

16
Chromosome mutation

Heritable changes in the genetic material can take place at the level of the chromosomes as well as at the level of individual genes. These **chromosome mutations** are gross changes in the genome: they can involve the *number* or the *structure* of chromosomes. In eukaryotes the term **genome** refers to *the sum total of genes in the basic set of chromosomes.*

Change in chromosome number

There are two sources of variation in the number of chromosomes: euploidy and aneuploidy. We will deal with euploidy first.

Euploidy

Euploidy is *variation in the number of whole sets of chromosomes.* We regard **diploids**, with two sets of chromosomes in their nuclei, as being the normal form in eukaryotes. **Monoploids** (or **haploids**) have only one set. They develop from unfertilised eggs and are very rare: we will not consider them any further (except in relation to parthenogenesis, Chapter 18). *Organisms with three or more complete sets of chromosomes* are known as **polyploids**. Polyploids are widespread and we must consider them in detail. The first thing to understand is how to represent their chromosome numbers.

Polyploidy

Chromosome numbers in polyploids   The convention for writing chromosome numbers, as explained in Chapter 2, is as follows:

x = the **basic number** of different chromosomes in a haploid set;
n = the number in the gametes, i.e. the **gametic number**;
$2n$ = the number in the zygote, i.e. the **zygotic number**.

In a diploid of course $x = n$, and we write the chromosome number as $2n = 2x = 20$ (e.g. maize). The $2n$ indicates that we are giving the zygotic number, and the $2x$ tells us that it is a diploid with *two sets* of 10 chromosomes in its somatic cells. To denote the level of polyploidy we simply use the appropriate value of x:

$$2n = 2x = 14 \quad \text{diploid}$$
$$2n = 3x = 21 \quad \text{triploid}$$
$$2n = 4x = 28 \quad \text{tetraploid}$$
$$2n = 5x = 35 \quad \text{pentaploid}$$
$$2n = 6x = 42 \quad \text{hexaploid}$$

polyploid series based on $x = 7$

We can only write the chromosome number in this precise way when we know that we are dealing with a polyploid, and when we are certain of its basic number. Quite often we don't know whether we are dealing with a polyploid or not, and then we give the zygotic number as (for example) $2n = 42$ and the gametic number as $n = 21$. The x value is unknown.

Polyploidy is rare among animals. For this reason their chromosome numbers are often written as $2n = 46$ (e.g. man) and it is taken for granted that they are diploid.

Occurrence of polyploidy   Polyploidy, by and large, is restricted to the plant kingdom. It is very rare in animals, except for some examples which will be mentioned in Chapter 18. The reasons for this disparity are not really known. Complications with the sex-determining chromosomes is one possibility; another is some kind of physiological disturbance. Where polyploids do arise in animals they usually fail to develop, or else they are aborted as in mammals.

Among flowering plants (angiosperms), on the other hand, polyploidy is common. Thirty five per cent of angiosperms are either polyploid species, or they are species which have polyploid *races* as well as diploids. Most of what we have to say about polyploids therefore refers to plants. A well-known example is the cultivated potato, *Solanum tuberosum* $(2n = 4x = 48)$. Its chromosomes are shown in Fig. 16.1, together with those of a wild diploid relative.

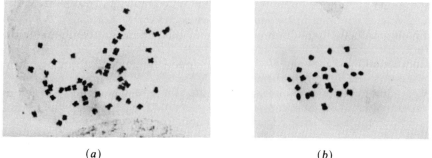

(a)   (b)

Fig. 16.1 (*a*) The cultivated potato, *Solanum tuberosum*, is a tetraploid with 48 chromosomes in its somatic cells $(2n = 4x = 48)$. Its wild relative, *S. brevidens* (*b*), is a diploid $(2n = 2x = 24)$. The basic number for the genus is $x = 12$. Chromosomes photographed at c-mitosis in cells of the root meristem, ×2000. (Photographs by A.K.)

Kinds of polyploids and their origin   There are basically two kinds of polyploids, autopolyploids and allopolyploids, and they arise in different ways.

Autopolyploids are *polyploids with more than two sets of chromosomes from within a single species.* They arise by spontaneous chromosome doubling. The most frequent way in which this happens is by spindle failure at meiosis, giving unreduced gametes (Fig. 16.2).

When an unreduced diploid gamete $(2x)$ fuses with a normal haploid one (x) a triploid $(3x)$ is produced. The union of two unreduced $2x$ gametes gives an autotetraploid. Triploids can also arise from crossing between diploids and tetraploids. Another source is spindle failure at mitosis: this gives direct doubling of the somatic chromosome number and leads to the production of a

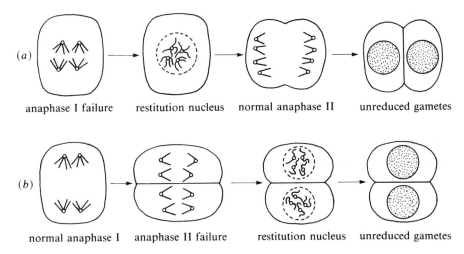

(a)   anaphase I failure   restitution nucleus   normal anaphase II   unreduced gametes

(b)   normal anaphase I   anaphase II failure   restitution nucleus   unreduced gametes

Fig. 16.2   Unreduced diploid gametes can result from a failure in the first (a) or second (b) meiotic division. In (a) the chromosomes do not separate properly at anaphase I, but instead form a common restitution nucleus. This nucleus then undergoes the second division to form two unreduced diploid gametes. In (b) anaphase II is abnormal. The chromatids do not separate but form two restitution nuclei which again give two diploid gametes.

polyploid cell. Derivatives of the doubled up cell, be they $4x$, $6x$ or $8x$, may then form a polyploid sector or branch in an otherwise $2x$, $3x$ or $4x$ plant. Autopolyploids have chromosome sets that are all homologous with one another. In the case of a tetraploid, with a basic number of $x = 3$, we may represent them as in Fig. 16.3.

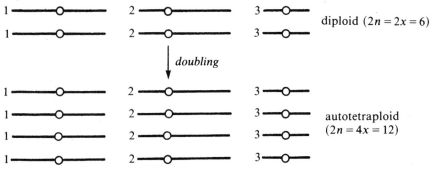

diploid $(2n = 2x = 6)$

doubling

autotetraploid $(2n = 4x = 12)$

Fig. 16.3

Examples of autopolyploids are the celandine (*Ranunculus ficaria*, $2n = 4x = 32$, $2n = 5x = 40$ and $2n = 6x = 48$); the hyacinth (*Hyacinthus orientalis*, $2n = 3x = 24$ and $2n = 4x = 32$) and some varieties of apple, e.g. Cox's Orange (*Malus pumila*, $2n = 3x = 51$).

Allopolyploids are *polyploids with multiple sets of chromosomes from more than one species.* The simplest way in which they can arise is by doubling in a hybrid between two related diploids. Another way is simply by crossing between two species that are already polyploid—say two tetraploids—in which case there is no need for doubling. They may also derive from hybridisation involving more than two species (see Box 16.2). Allopolyploids are different from autopolyploids in that the chromosome sets contributed by the two (or more) parent species are *not* completely homologous (Fig. 16.4).

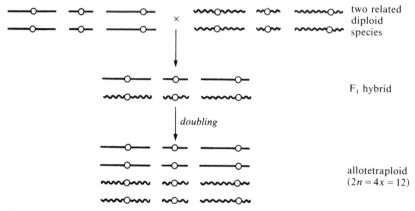

Fig. 16.4

Examples of allopolyploids are to be found in the hempnettle, *Galeopsis tetrahit* $(2n = 4x = 32,\ G.\ pubescens \times G.\ speciosa)$; the tobacco, *Nicotiana tabacum* $(2n = 4x = 48,\ N.\ sylvestris \times N.\ tomentosiformis)$ and in *Primula kewensis* $(2n = 4x = 36,\ P.\ floribunda \times P.\ verticillata)$.

Experimental production of polyploids   Chromosome doubling can be induced experimentally by the use of chemicals which interfere with spindle formation. Colchicine is the most widely used substance for this purpose. It disrupts the organisation of the spindle microtubules, but when it is withdrawn the cells can again divide normally. Plant breeders immerse young seedlings in a dilute aqueous solution of colchicine to make new polyploid varieties of crop plants.

Consequences and significance of polyploidy   Polyploids are important in nature because of the way in which they affect (1) physiology and development, (2) genetic variation and (3) meiosis and reproduction. For these reasons they differ in their adaptation from diploids and may be able to colonise and to occupy new and different habitats to those of their diploid relatives. In the case of allopolyploids, entirely new species may be created.

Polyploids have larger nuclei than their diploid progenitors, and this in turn gives rise to an increase in cell size and an altered pattern of development and morphology. Leaves tend to be larger and thicker, the growth rate is slower, and there may be physiological differences in response to certain environmental factors (temperature, rainfall). These **gigas** effects are not invariable and they are generally much more pronounced in newly formed polyploids than in those which are well established.

The consequences in terms of genetic variation differ between autopolyploids and allopolyploids. In an autotetraploid, for instance, which has two allelic forms of a gene at a given locus (**A** and **a**), we can have five genotypes in the population—**AAAA, AAAa, AAaa, Aaaa, aaaa**. In a diploid we have only three (**AA, Aa, aa**). One result is that the polyploid population will contain a much higher proportion of heterozygotes (**AAAa, AAaa, Aaaa**) and a greatly reduced fraction of homozygotes (**AAAA, aaaa**). We will not attempt to explain why this is so: suffice it to say that the pattern of genetic variation will be different from that in a diploid. There are also complications in the pairing and distribution of chromosomes at meiosis. The details are given in Box 16.1. The point is that autopolyploids have problems in distributing their chromosomes in 'balanced' sets into the gametes. This is especially true in the odd-numbered ones (3x, 5x, etc.). Their gametes are often inviable (i.e. fail to develop) and the plants do not set seed properly. It is for this reason that autopolyploids have evolved the various methods for asexual reproduction (bulbs, corms, tubers, etc.) that will be described in Chapter 18. Even-numbered polyploids have much less of a problem, as explained in Fig. 16.5.

Box 16.1   Chromosome pairing in autopolyploids

Polyploidy affects the pairing and distribution of chromosomes at meiosis. In a diploid the homologous partners form **bivalents**. In an autotriploid they associate in threes (**trivalents**), and in an autotetraploid they pair in fours (**quadrivalents**). Triploids cannot distribute their chromosomes equally at anaphase I and the gametes contain unbalanced numbers of the different members of the basic set. Consequently, they are sterile and fail to set any seeds. Tetraploids can manage much better, but they also have problems if pairing takes the form of a trivalent and univalent instead of a quadrivalent.

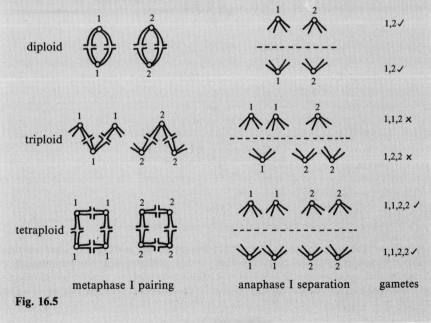

metaphase I pairing   anaphase I separation   gametes

Fig. 16.5

Allopolyploids combine the genetic qualities of two different species into a new fertile hybrid. The F_1 from diploid parents is usually sterile because the chromosomes from the two parent species are not homologous and cannot pair. After doubling, each chromosome has an identical partner. The chromosomes can then associate in pairs and the new species can behave as a diploid at meiosis.

The Russian geneticist Karpechenko (1928) first explained how chromosome doubling in a species hybrid could produce a fertile allopolyploid. The cross he made was between the radish (*Raphanus sativus*, $2n = 2x = 18$) and the cabbage (*Brassica oleracea*, $2n = 2x = 18$)—Fig. 16.6. The most important example of a natural allopolyploid is the bread wheat, *Triticum aestivum* ($2n = 6x = 42$), which was taken into cultivation by man more than 6000 years ago (Box 16.2).

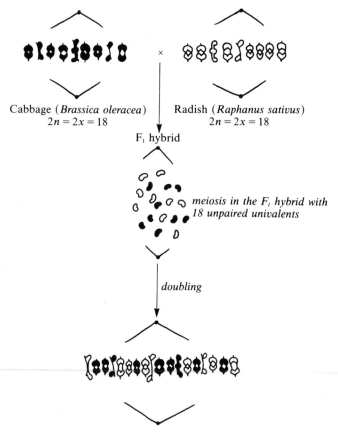

Cabbage (*Brassica oleracea*)
$2n = 2x = 18$

Radish (*Raphanus sativus*)
$2n = 2x = 18$

F_1 hybrid

meiosis in the F_1 hybrid with 18 unpaired univalents

doubling

fertile allotetraploid with 18 bivalents

Fig. 16.6 Chromosome pairing at MI of meiosis in the cabbage (*B. oleracea*), the radish (*R. sativus*), their F_1 diploid hybrid and the allotetraploid. The parental sets of chromosomes are structurally different and cannot pair together in the F_1. Meiosis fails and the diploid hybrid is sterile. After doubling, each chromosome has an identical partner and a fully fertile allotetraploid is formed in which the chromosomes pair as bivalents.

Box 16.2   Bread wheat—origin of an allohexaploid

The bread wheat, *Triticum aestivum*, is the most widely cultivated plant in the world. It is an allohexaploid, $2n = 6x = 42$ (Fig. 16.8). Three diploid wild grasses are involved in its origin, each one contributing a different genome. The three genomes, i.e. the three different basic sets of chromosomes, are represented as A, B and D. The source of the B genome is unknown. The A genome came from *T. monococcum* which hybridised naturally with the B genome donor to give the wild allotetraploid *T. turgidum*. A cultivated variety of the tetraploid (var. dicoccum) crossed naturally with wild *T. tauschii* (growing as a weed) to give the cultivated allohexaploid, *T. aestivum*. Modern varieties of bread wheat have their origin in these natural hybridisation events which took place more than 6000 years ago:

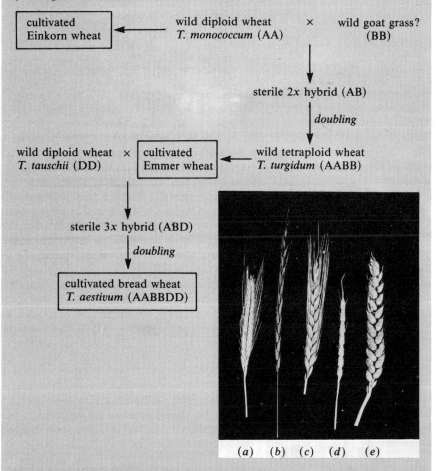

Fig. 16.7   The origin of bread wheat. The wild diploid wheat *Triticum monococcum* (*a*) hybridised naturally with a wild goat grass (*b*) to give the wild tetraploid species *T. turgidum*. A cultivated form of *T. turgidum* (*c*) then crossed naturally with another wild diploid speces *T. tauschii* (*d*) to give the hexaploid bread wheat (*e*) which forms the basis of all our present-day varieties of *T. aestivum*.

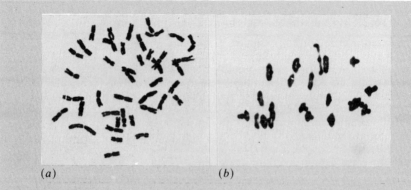

Fig. 16.8 Chromosomes of bread wheat. (*a*) C-mitosis in a root meristem cell. (*b*) At MI of meiosis the chromosomes within each of the three genomes come together in pairs to form 21 bivalents. This ensures their perfect segregation and complete self-fertility. (Photographs by A.K.)

The distinction between autopolyploids and allopolyploids We have described chromosome pairing behaviour in auto- and allopolyploids in simple terms and have left out many of the details and complications. In reality the difference between these two groups is not usually as clear-cut as we have suggested. The distinction may be blurred because the parent species that hybridise together to form allopolyploids vary widely in their degree of relatedness. In the case of *Raphanus sativus* × *Brassica oleracea* the hybrid is **intergeneric** (i.e. the two species are not even in the same genus), and the two chromosome sets in the F_1 are widely divergent in their structure and genetic organisation. Hybrids between two closely related species from the same genus (e.g. the ryegrasses *Lolium perenne* × *L. temulentum*) have chromosome sets with a much greater degree of **homology**. They can sometimes form bivalents at meiosis in the F_1, and a mixture of both bivalents and multivalents at the allopolyploid level. Their chromosome pairing is intermediate between that of an autopolyploid and an extreme form of allopolyploid. Another complication arises from the fact that pairing behaviour can be controlled by single genes, as well as by structural divergence between different complements—which is the case in wheat.

Aneuploidy

Aneuploidy refers to *change in number involving only part of a chromosome set.* One or more whole chromosomes may be absent from or in addition to the diploid or polyploid complement. The absence of a chromosome in a diploid is denoted as $2n = 2x - 1$, and the addition by $2n = 2x + 1$.

Origin of aneuploidy The irregular distribution of chromosomes at meiosis in polyploids, particularly the odd-numbered ones, is one source of aneuploidy. This applies mainly to plants, since polyploidy is rare in animals. Aneuploids may also be produced by a process of **non-disjunction**. This is a mistake in chromosome separation in which a pair of chromosomes (or chromatids) pass to the same pole of a cell instead of to opposite poles. It can happen at mitosis or meiosis. It is more likely to give aneuploid progeny when it occurs at meiosis (Fig. 16.9). This is the way in which aneuploids arise in animals, including man.

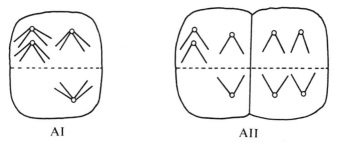

AI   AII

Fig. 16.9 Diagram showing how aneuploidy arises by non-disjunction at AI or AII of meiosis. In the AI cell a pair of chromosomes is also shown disjoining normally. At AII both cells are shown—one is normal and the other has chromatid non-disjunction in one of its chromosomes.

Consequences of aneuploidy Aneuploidy causes unbalance in the chromosome complement and nearly always results in an abnormal phenotype. An example is shown in Fig. 16.10. In plants the effects are much less severe in polyploids than in diploids. The degree of abnormality depends upon which *particular* chromosome is missing or additional, and in general an extra chromosome is less deleterious than a missing one.

(a)   (b)   (c)

Fig. 16.10 Aneuploidy in perennial ryegrass (*Lolium perenne*). (a) Normal diploid plant ($2n = 2x = 14$). (b) Phenotypically abnormal trisomic plant carrying an extra chromosome in all its cells ($2n = 2x = 14 + 1$). (c) Chromosome pairing at metaphase I of meiosis in the aneuploid: there are 6 bivalents and a *trivalent* of three homologous chromosomes (arrowed).

Aneuploidy in man The commonest aneuploids in man are those involving the sex chromosomes. The unbalance results in abnormalities in development but it is not generally lethal. A well known example is **Turner's syndrome** (XO), in which there is only one X and no Y. The affected individuals are female, but they are sexually underdeveloped and are sterile. Aneuploidy in males includes **Klinefelter's syndrome** (XXY and XXXY). In these cases the persons concerned have additional Xs but they develop as males due to the presence of the Y chromosome. They are also sterile and have effeminate tendencies.

Another category which occurs in about 1 in 1000 births, is XYY. These males are usually fertile and may be normal in phenotype. There appears to be a significantly higher proportion of them among socially deviant males than there is within the population in general.

Down's syndrome (mongolism) is the best known of the autosomal aneuploids. One form is due to trisomy (three copies) for chromosome 21 (Fig. 16.11). The phenotype is characterised by mental retardation and certain distinctive physical features. An important aspect of Down's syndrome is its incidence in relation to the age of the mother (Fig. 16.11). Women over the age of 40 who become pregnant are advised to have the chromosomes of the foetus examined. This can be done because a few cells get sloughed off the foetus into the amniotic fluid. A sample of this fluid containing the cells is then removed through the wall of the uterus (with a syringe) and the cells multiplied up in tissue culture for chromosome analysis. This method of sampling and analysis of the amniotic fluid is known as **amniocentesis**.

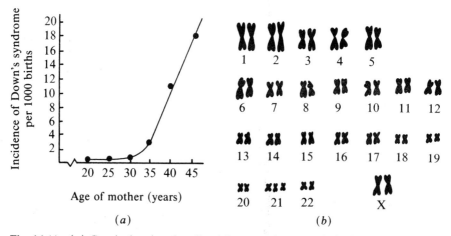

Fig. 16.11 (*a*) Graph showing the effect of maternal age on the incidence of children with Down's syndrome. (*b*) Karyotype of a female with Down's syndrome due to trisomy for chromosome 21.

Trisomy is also known for chromosome 18 (**Edward's syndrome**) and 13 (**Patau's syndrome**). Aneuploidy for any other chromosomes in the complement, aside from 21, 18, 13, X and Y, is lethal. It is estimated that out of every 1 000 000 conceptions in man there are 150 000 spontaneous abortions, of which 75 000 are due to some kind of chromosome abnormality (aneuploidy, polyploidy and structural change). Out of the remaining 850 000 live births some 833 000 survive and 5165 of these carry a chromosome mutation. The frequency of chromosome mutation in the human population is therefore of the order of 0.61% of live births, i.e. more than 1 in 600.

Change in chromosome structure

Variation in chromosome structure results from breakage and from errors in crossing over. Breakages may, or may not, be followed by rejoining of the broken ends. As these changes are the result of random events, it is unlikely that both members of a pair of homologous chromosomes in diploid cells will be affected in the same way at the same time. Consequently the *structural* changes arise in heterozygous form. We recognise four main categories: deletions, duplications, inversions and translocations.

Deletions

A **deletion** (or **deficiency**) refers to *the loss of a chromosome segment* (Fig. 16.12).

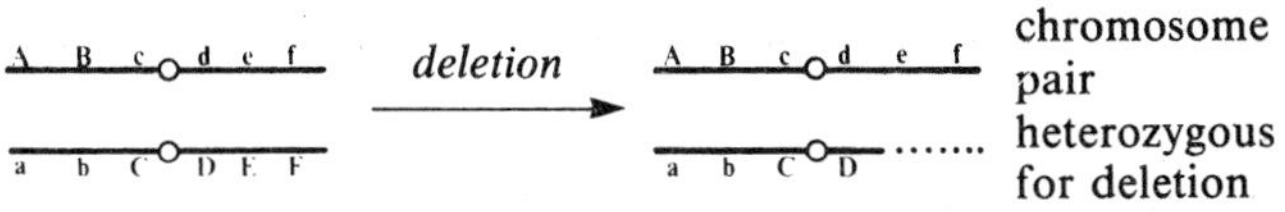

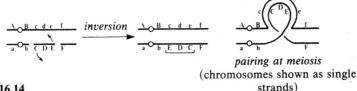

Fig. 16.12

Deletions are usually lethal even in the heterozygous condition. As homozygotes they will certainly be lethal if the genes which are lost are concerned with some essential function. A well known example of a heterozygous deletion in man is the '*Cri-du-chat*' syndrome, which is due to the loss of a segment from chromosome 5. It causes mental retardation, abnormalities of the face and head, and a characteristic high-pitched cry which resembles that of a cat in distress.

Duplications

A **duplication** occurs when *a chromosome segment is present more than twice in a diploid* (Fig. 16.13).

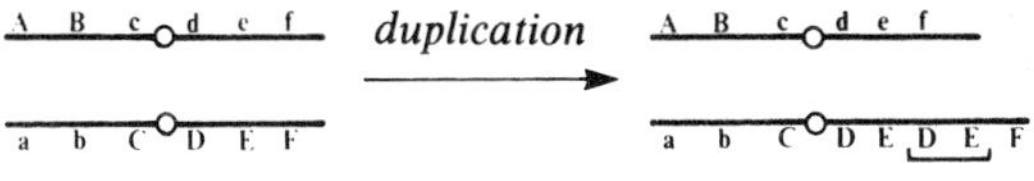

Fig. 16.13

Duplications can arise from errors in crossing over. They affect the phenotype because of the altered dosage. The effect depends upon the particular segment concerned, but clearly duplications are less harmful than deletions as there is no loss of genetic material. Duplications are important in evolution. When more than one copy of a gene is present, the redundant one is free to mutate and to evolve a new function. The α and β haemoglobin genes (p. 193) are thought to have arisen in this way.

Inversions

An **inversion** is *the reversal of the gene order which may result when two breaks occur in the same chromosome* (Fig. 16.14).

pairing at meiosis
(chromosomes shown as single strands)

Fig. 16.14

There is no loss or duplication of genetic material in an inversion, but there is a problem at meiosis to do with pairing in the heterozygote. The two homologues cannot align themselves side by side in the same way that they would normally do at prophase I, because the gene order is reversed in one of them. They overcome this difficulty by forming an **inverse pairing loop** (Fig. 16.14). If crossing over takes place within the loop, the chromosomes have difficulty in separating and half of the gametes which are produced are inviable. That is why inversions are important—they prevent recombination taking place between adaptive combinations of linked genes within the inverted region. Crossing over can take place elsewhere within the bivalent, of course. There is no difficulty at meiosis in the inversion homozygote (i.e. where the same inversion is present in both homologous chromosomes), and no problem at mitosis in either heterozygotes or homozygotes.

Translocations

In a **translocation** *a segment is transferred from one chromosome to another, non-homologous one.* If two chromosomes exchange segments the translocation is then sometimes called an **interchange** (Fig. 16.15).

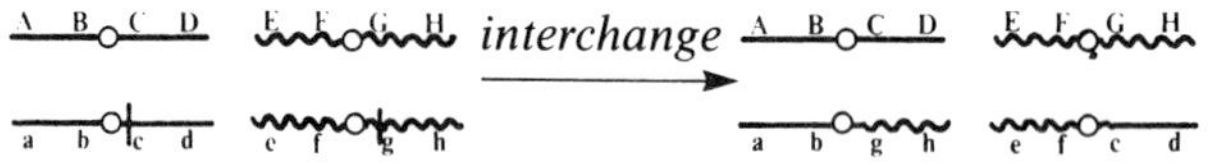

Fig. 16.15

As with inversions these translocations do not affect mitosis in any way, but they cause complications at meiosis as heterozygotes. The reason is that two pairs of chromosomes now have some homologous parts in common, and four chromosomes associate together at prophase I. We will not attempt to explain the complexities of chromosome pairing and segregation in an interchange heterozygote. We will simply say that there may be infertility due to the production of inviable gametes, and that interchanges are important because they link together genes in different pairs of chromosomes and prevent their independent segregation. As with inversions they therefore regulate the pattern of genetic variation in natural populations.

Summary

Chromosome mutations cause major changes in the quantity and quality of the genetic material in the nucleus. Numerical changes involving whole sets of chromosomes affect all aspects of development, genetic variation, meiosis and reproduction. They may even give rise to entirely new species, including several which are of use to man. Changes in part of a set (aneuploidy) are generally deleterious because they upset the genic balance between different chromosomes within the complement. Aneuploidy is a major problem in the human race. Structural changes are important in terms of the evolution of chromosomes, and for the way in which they can regulate the pattern of genetic variation in natural populations—although this aspect is not discussed in any detail.

Further reading

Darlington, C. D. (1956), *Chromosome Botany*, George Allen and Unwin.
Stebbins, G. L. (1971), *Chromosome Evolution in Higher Plants*, Edward Arnold.

Questions on Chapters 15 and 16

1 Write a detailed account of the modern explanation of how mutations occur.
(W.J.E.C. Biol. S., 1981)

2 Describe gene and chromosome mutations and discuss their relative importance in the evolution of new forms of plants and animals.   *(W.J.E.C. Biol. S., 1982)*

3 (*a*) What are 'genes' and how are they involved in inheritance?
 (*b*) Explain what is meant by the terms: 'polyploidy', 'mutation' and 'gene linkage'.
(C. Bot., 1981)

4 Discuss the importance of polyploidy with special reference to hybrid sterility. Why do you think polyploids are so common among many of the higher plants and yet so rare among the higher animals?   *(O.L.E. Bot., 1979)*

5 Write an essay on genetic mutations. Discuss the importance of mutation in human medicine.   *(W.J.E.C. Biol., 1978)*

6 The plant genus *Brassica* (Cruciferae) contains a number of species from which some of our commonest vegetable and fodder plants, e.g. cabbage, swede and rape have evolved. Hybridisation between the turnip (*Brassica rapa*, $2n = 20$) and the black mustard (*Brassica nigra*, $2n = 16$) has produced the brown mustard (*Brassica juncea*, $2n = 36$).
 (*a*) How are the hybrids produced experimentally?
 (*b*) What chromosomal change occurred during the formation of the brown mustard hybrid, and what term is used to describe such a hybrid?
 (*c*) Why are plants of brown mustard able to produce fertile seed?
 (*d*) A black mustard plant with 32 chromosomes has been produced but it has proved to be almost totally sterile. Explain why this is so.
 (*e*) Suggest three reasons why closely related species rarely form hybrids in nature.
(O.L.E. Biol., 1982)

7 (*a*) What is meant by 'mutation'?
 (*b*) State two agents, other than ionising radiation, which induce mutations.
 (*c*) Give two reasons why micro-organisms such as bacteria are frequently used in the study of mutation.
 (*d*) Why is mutation considered to be an important factor in the evolution of new species?

17
Regulation of gene action

Multicellular organisms arise from single fertilised egg cells. In the course of development all of the cells which comprise an individual are produced as a result of mitosis. Mitosis is the division of the nucleus which distributes identical copies of the genetic material to daughter cells (Chapter 2). We have seen how the process works at the molecular level in Chapter 13. It has to do with the semi-conservative replication of DNA and the way in which the products of self-replication are separated at mitosis when the chromosomes divide. In terms of phenotype, however, we also know that the cells of an organism are not identical. They display a wide variety of forms and differ greatly from one another in their enzymes and other kinds of proteins (e.g. liver, muscle and skin cells). The term **differentiation** is used to describe *the formation of distinctly different cell types from a single-celled zygote.*

How is it that the zygote, which develops by mitosis, can differentiate into a variety of cell types that differ so widely in their form and in their physiology? The answer, evidently, is in the selective utilisation of their genetic information. Different genes are used in different cells and at different stages of development. The term **development** refers to *the organisation of cells within the tissues and organs of a multicellular organism.* We will not concern ourselves here either with the processes of development itself, or with what is known about the genetic control of development. This is one of the most difficult and least well understood areas of biology and one which poses the greatest challenge at the present time.

In this chapter we will simply confine ourselves to some examples which show that different genes are active in different tissues, and at different times in development, and to discussing some of the mechanisms by which their activities are regulated.

Totipotency

Totipotency is *the capacity of differentiated somatic cells to retain their potential to produce an entire organism.* Totipotency can be demonstrated in a number of plant and animal species and this provides the evidence that somatic cells retain their full complement of genes during development and differentiation. In 1952 F. C. Steward showed that entire carrot plants could be regenerated from single cells of the leaf grown in culture medium. The techniques of plant tissue culture have now become so well developed that the regeneration of whole plants from cultured single cells, or from small pieces of tissue, can be routinely achieved in a wide variety of species.

Totipotency is more difficult to demonstrate in animals, but it has been done by nuclear transplantation experiments in amphibians. The nucleus is removed from an unfertilised egg of a frog and then replaced, by microsurgery, with a diploid somatic cell nucleus of a tadpole. In a proportion of such transfers the eggs develop into adult frogs, showing that the nuclei of the tadpole's cells have retained their capacity to produce an entire organism. To date there are no known cases where an animal has been regenerated from a single *somatic* cell.

Differential activity of genes

One of the most obvious ways to show that different genes are active in different types of cell, and at different stages in development, is to look directly at the products of their action, i.e. at proteins. Haemoglobin in man is a good example because it has been so intensively studied.

Haemoglobin

Haemoglobin is a complex protein comprising four polypeptide chains, as already explained in Chapter 14. The genes which code for these polypeptides are present in all the cells of the body, but they are only active in one particular cell type—namely the **erythroblasts**. Erythroblasts are produced from stem cells in the bone marrow, although neither the stem cells themselves, nor the erythroblasts actually contain any haemoglobin. The genes are transcribed in the erythroblasts where they synthesise their mRNAs. These cells then give rise to **reticulocytes** which lose their nuclei, and it is only after this that the messenger RNAs are translated to produce the haemoglobin in the mature red blood cell (**erythrocyte**) (Fig. 17.1).

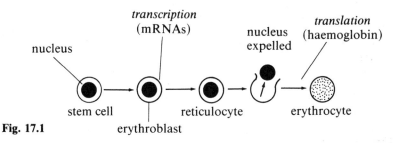

transcription
(mRNAs)

translation
(haemoglobin)

nucleus

nucleus
expelled

stem cell   reticulocyte   erythrocyte

Fig. 17.1   erythroblast

There are in fact seven different genes involved in the production of haemoglobin, each one coding for a different polypeptide chain. One of these genes, specifying the alpha (α) polypeptide, is active throughout life. The others are regulated to function only at certain times in development. The sequence of appearance of the seven polypeptides, and the way in which they combine together to produce four different types of haemoglobin, is illustrated in Fig. 17.2. Two of the genes, coding for the zeta (ζ) and epsilon (ε) polypeptides, produce **embryonic haemoglobin**. They are active during the first three months of life and then they are switched off. The two kinds of gamma chains ($^A\gamma$, $^G\gamma$) appear in the early embryo and reach their peak level between the third month of development and birth. They combine with the α polypeptides to form **foetal haemoglobin**. Beta (β) and delta (δ) chains are produced from about the time of birth onwards, for the rest of life. In the 'adult' phase of development, i.e. from about six months after birth onwards, 98% of all haemoglobin is the **adult haemoglobin (HbA)** in which each molecule is comprised of two α and two β polypeptides (Chapter 14). The remaining 2% is **minor adult haemoglobin (HbA2)**, made from two α and two δ chains.

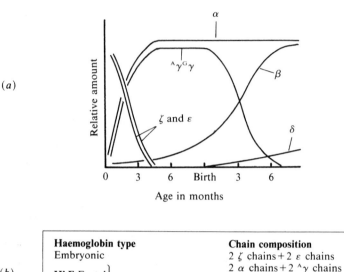

(a)

(b)

Haemoglobin type	Chain composition
Embryonic	2 ζ chains + 2 ε chains
HbF Foetal }	2 α chains + 2 $^A\gamma$ chains
	2 α chains + 2 $^G\gamma$ chains
HbA Adult	2 α chains + 2 β chains
HbA2 Minor adult	2 α chains + 2 δ chains

Fig. 17.2 (a) Sequence of appearance of the 7 different kinds of polypeptide chains which make up the four types of human haemoglobin. Each of the chains is coded for by a different gene. The sequence of appearance of the chains reflects the differential activity of the genes and shows how different genes may be switched 'on' and 'off' at specific times in development.

(b) The four types of human haemoglobin are made up from various combinations of two identical pairs of chains. **HbA adult haemoglobin** (2 α and 2 β chains) accounts for 98% of all haemoglobin in persons who are more than 6 months of age.

Polytene chromosomes

The salivary gland cells and various other secretory-type tissues in *Drosophila*, and in other dipteran flies, contain polytene chromosomes. **Polytene chromosomes** are *giant chromosomes which arise by endomitosis, in which the chromosomes replicate repeatedly without any separation of their chromatids.* These chromosomes are unusual in that they contain as many as 1000 chromatids. This makes them huge in size and they can be seen quite easily under the microscope even though they are in the interphase stage of the cell cycle. What is more, the homologous partners are paired together in intimate somatic association throughout their length. The over-replication, and close parallel alignment of chromatids, grossly amplifies the structure of the chromosomes and makes visible a remarkable amount of detail of their organisation. When they are stained with suitable dyes it is possible to see numerous dark bands arranged in a characteristic pattern along their length and interspersed with lightly-stained interband regions (Fig. 17.3*a*). In *Drosophila* the number of bands corresponds roughly with the number of gene loci. It cannot be said though that we can see the genes, because each of the chromatids within a band contains about 25 times more DNA than the amount required to make a single gene—so the genetic organisation of the bands is not fully understood.

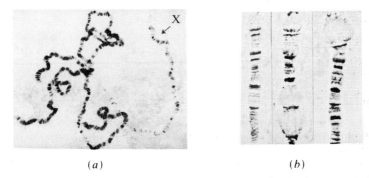

(*a*) (*b*)

Fig. 17.3 (*a*) Polytene chromosomes from a salivary gland cell of a male *Drosophila* ($2n = 2x = 6 + XY$). The homologous chromosomes are paired together and also fused in the regions of their centromeres (centre of photo). Chromosome 4 is very tiny and the Y chromosome is also so reduced in size as to be indistinguishable in the centre. The single X is unpaired and is clearly visible as a much more slender chromosome than the two arms of chromosomes 2 and 3 which make up the rest of the picture. As well as the salivary glands these polytene chromosomes can also be found in various other secretory tissues such as the epithelial cells of the rectum, the gut and the Malpighian tubules. They occur in many other dipteran insects and also in some plant tissues, e.g. the suspensor cells of the *Phaseolus* (runner bean) embryo. (Photograph by Mared Breese.)

(*b*) Higher magnification of a region of chromosome arm 3L of *Drosophila*, from larvae of three different ages. Note the presence of chromosome puffs and the change in the pattern of their formation during larval development. (Photograph courtesy of Dr Michael Ashburner, Genetics Department, University of Cambridge.)

These polytene chromosomes, because they are so large, offer a useful opportunity for studying gene action cytologically, i.e. by looking at them down the microscope. During development of the fly larvae, some of the bands undergo drastic changes in their morphology. They swell up and appear to unfold and spill out their DNA strands (Fig. 17.3*b*). These swollen-up bands are called **chromosome puffs**, or when they are very extensive **Balbiani rings**. When salivary gland chromosomes are incubated with radioactive uracil (one of the bases in RNA) they take up the 'label' which may then be seen, by autoradiography, to be localised in the puffs. This kind of experiment confirms that the puffs represent sites of genetic activity and that in these places the spilled-out DNA strands are being actively transcribed into mRNA, which has incorporated the labelled uracil bases.

Wolfgang Beerman has carried out detailed studies on the patterns of puffing in the polytene chromosomes of the midge *Chironomus*. He found that relatively few bands were actually forming puffs in any one tissue, but that the patterns of puffing along the chromosomes were characteristic of particular tissues (salivary glands, rectum, gut, Malpighian tubules) and these patterns changed at different stages of larval development. These studies demonstrate quite clearly the way in which different genes are active in different tissues and at different times in development.

Mechanisms of gene regulation

It is evident from what has been said above that the activity of genes is under regulatory control. During development the genetic information in the nucleus is used in a selective way so that the genes are expressed when and where they are needed, so fulfilling their specific functions. We will now consider some of the mechanisms responsible for gene regulation. As we will see, they are not all concerned with simply switching genes 'on' and 'off'; neither are they all involved in developmental processes. Certain genes are regulated in the course of normal physiological activity to do with the routine working of the cells and of the organism, e.g. light-regulated genes which control photosynthesis in plants.

Much of what we know about the control of gene action comes in fact from the study of prokaryotes, rather than from multicellular eukaryotes. This is because of their simpler organisation and the much greater facility which they offer for genetic experimentation (Chapter 12).

Prokaryotes

One of the best known examples of gene regulation in prokaryotes is that concerning the utilisation of lactose by the bacterium *E. coli*. The system was worked out by the French scientists François Jacob and Jacques Monod in the 1960s.

E. coli is able to utilise lactose as a source of energy. To do so it uses three enzymes:

1. **β-galactosidase permease**, which is involved in the transport of lactose into the cell and is coded for by a gene designated **y**;
2. **β-galactosidase**, which splits lactose into its two components—galactose and glucose—and which is determined by gene **z**;
3. An **acetylase** enzyme specified by gene **a** which is also concerned with lactose metabolism.

The three enzymes are **inducible**, that is to say they are only synthesised within the cell when their substrate (lactose) is present in the growth medium. If cells are transferred to a medium which is lacking in lactose then their synthesis stops within 2–3 minutes and only minute traces of them are then present. The three structural genes which code for the enzymes are therefore regulated in some way so that they can respond to the presence or absence of lactose and only make their enzyme products when they are actually required by the cell. By using mutations which affected the activity of the three enzymes Jacob and Monod were able to map the location of their genes in the bacterial chromosomes. The genes were found to be **contiguous**, i.e. located side-by-side, **z y a**, as shown in Fig. 17.4.

In addition to the **z**, **y** and **a** genes there are two other genetic elements involved in lactose utilisation which do not code for any enzymes. These are the **regulatory** gene i and **operator** site o, which serve to control the action of the three structural genes. Mutations in either the regulator gene or the operator causes the structural genes to behave **constitutively**, i.e. to make their enzymes all the time irrespective of whether there is any lactose present or not. When the regulation is lost in this way it affects all three structural genes at the same time, showing that they are normally under **co-ordinate control** by the i and o regulators. The operator site is located next to the z gene in the chromosome. The regulator i is separated from o by a promoter region (**p**) which is the site of attachment of the RNA polymerase enzyme which catalyses transcription (Fig. 17.4).

(*a*) Repressed: lactose absent, genes 'switched off'

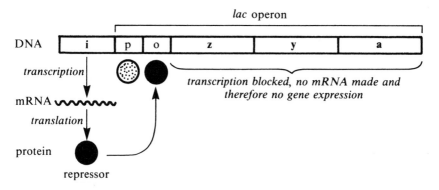

(*b*) Induced: lactose present, genes 'switched on'

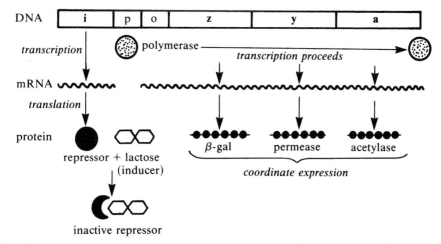

Fig. 17.4   Diagram showing the *lac* operon in its repressed (*a*) and induced (*b*) states. In the repressed state the repressor protein binds to the operator site and prevents transcription. The genes z, y and a are then switched off. When lactose (the inducer) is present, it alters the shape of the repressor which can then no longer bind to the operator. The genes are then transcribed. (i = regulator gene; p = promoter site; o = operator site; z, y and a = structural genes.)

Jacob and Monod use the term **operon** to describe *a group of co-ordinately expressed and adjacent structural genes, together with their operator and promoter sites.* They proposed the scheme described in Fig. 17.4 to explain the mechanism of control in the lactose operon (*lac* operon) in *E. coli.* In essence the system is quite simple. When lactose is absent, a **repressor** protein made by gene **i** binds to the operator and blocks transcription, so that the structural genes are **repressed** or 'switched off'. When lactose is present it binds to the repressor molecules and prevents them from blocking the operator—transcription can then proceed and the genes are 'switched on', i.e. their activity is induced by the lactose. Repressor molecules are usually present in the cell in quite a low concentration. Mutations in gene **i** or site **o** cause the structural genes to behave constitutively, as described above, because they either affect the repressor itself ($i^+ \rightarrow i^-$) or else they prevent it binding to the operator site ($o^+ \rightarrow o^c$). As Fig. 17.4 shows, the co-ordinate expression of the three structural genes comes about because of the way in which they are all adjacent to one another and all transcribed as one long strand of mRNA. During translation, of course, the stop codons between the genes result in three separate polypeptide chains (Chapter 14).

There are other examples of operons which are known in bacteria, suggesting that this is a common mode of gene regulation in prokaryotes, although the regulator does not always function in exactly the same way.

Eukaryotes

There are several important differences in the organisation of prokaryotic and eukaryotic genomes which affect the organisation of their DNA and the way in which genes are transcribed and translated. Operons of the type seen in prokaryotes, are not found in eukaryotes.

Although genes with related function are often **clustered**, the control of their expression may extend over several different clusters, or even over genes on different chromosomes. The genomes of eukaryotes are much larger and contain many kinds of repetitive DNA sequences whose function is unknown (p. 202). Eukaryotic genes can also be 'split' by introns (p. 202) and their DNA is bound up with histones in the form of chromatin (Chapter 13). In view of this complexity it is not surprising that gene regulation is much less well understood in eukaryotes.

In eukaryotes the transcription and translation of genes are not coupled in the way that they are in prokaryotes (p. 199). Transcription takes place within the nucleus and translation occurs later, on the ribosomes in the cytoplasm. Regulation of genes can therefore take place at either of these stages in the flow of information.

Regulation at the level of transcription   There are more examples known of control at the level of transcription than at the level of translation. One such case is thought to operate during the formation of seed-storage proteins in some plant species, as for example in barley (*Hordeum vulgare*). There are several types of seed-storage proteins in barley and collectively they are referred to as the **hordeins**. The genes which code for them are active in the seeds only over a three-week period of development. During this time there are changes in the relative amounts of the three main kinds of hordein polypeptides being produced (Fig. 17.5). These changes are due to developmental control over the action of the genes and can be detected by similar patterns of change in their corresponding mRNAs. It is the close correspondence between the messenger RNAs and their products of translation which suggests that control is operating at the level of transcription. Once transcribed, all of the messengers are translated into protein. The mechanism of control is thought to be similar to that described for prokaryotes, only in this case the details are not known and the regulatory genes which have been identified are located on a different chromosome from the structural genes.

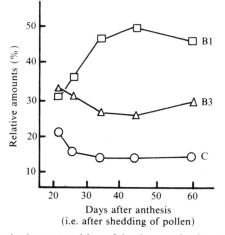

Fig. 17.5   Changes in the composition of the three main classes of seed storage proteins (B1, B3, C) in barley during seed development. Synthesis of the proteins begins two to three weeks after anthesis and continues until seed development is complete. The pattern of change in polypeptide composition is accompanied by a corresponding pattern of variation in mRNA transcripts. (Data from S. Rahman, Biochemistry Department, Rothamsted.)

Regulation at the translational level   Differences in protein synthesis cannot always be explained by changes in the proportion of different transcripts making up the population of mRNAs in a cell. In some instances we find cells which have identical complements of mRNAs but different kinds of proteins, because not all the mRNAs are being translated. This is regulation at the level of translation. Examples can be found in some animals where the unfertilised egg cell is packed with mRNA, which is then selectively translated when the zygote begins to develop after fertilisation.

In the surf clam (*Spisula solidissima*) there are striking differences in the proteins synthesised before and after fertilisation, but no differences in the respective kinds of mRNAs in the two phases of development. By extracting ribosomes from the unfertilised egg, and the young embryo, it is possible to examine the nature of the mRNAs which are actually being translated during these stages, because the mRNAs are associated with the extracted ribosomes. Such studies show that different mRNA transcripts are selected for translation before and after fertilisation. The way in which this translational control works is not fully understood, but there is some evidence that it may operate through the association of specific proteins with mRNA molecules.

Gene regulation at the level of chromosome organisation   The activity of eukaryotic genes can also be controlled by mechanisms which involve changes in the state of organisation of the chromatin within the chromosomes. One example is the system described in Chapter 8 where one of the two X chromosomes in female mammals is inactivated by a process of **heterochromatinisation**. The interphase X chromosome coils up tightly and is visible as the **Barr body**. In this condition its DNA is unavailable for transcription and all of the genes within the chromosome are thus 'switched off'. In this case the regulation is acting as a dosage-compensation device which allows for the fact that the male has only one X chromosome, whereas the female has two.

Gene amplification   As a final example of some of the many different ways in which eukaryotes can regulate their genes we will refer to **gene amplification**. This mechanism is well known in the oocytes of amphibians, such as the South African clawed toad *Xenopus laevis*, and it is an economical means of satisfying demand for a large quantity of gene product at certain times in development. In these oocytes there is a need to synthesise enormous quantities of ribosomes in a relatively short space of time, so that protein synthesis can proceed quickly as soon as fertilisation has taken place. This is achieved by amplification of the region of the chromosomes which carries the ribosomal RNA genes, i.e. the nucleolus organiser region. Additional copies are made (in excess of 1000) of this one small segment of one of the chromosomes. It happens by what is known as the **rolling circle mechanism** of DNA replication and one copy after another of the relevant rRNA sequences in the DNA are replicated and then released from the chromosome. Each one of the multiple copies in the nucleolus organiser region then proceeds to organise its own nucleolus and to transcribe its ribosomal RNA genes.

Summary

The action of genes is regulated. This regulation is necessary because in multicellular organisms there is a full complement of genes in every cell and not all of these genes are required to function all of the time. Their activities are controlled in various ways so that only those gene products which are required at a specific time in development are actually produced. Control is also needed to ensure that gene products needed for the routine physiology of the cell, and the organism, are also made at the appropriate time. The *lac* operon in *E. coli* is discussed in detail. Although *E. coli* is a unicellular organism, it does provide us with one of the best known examples of how gene action can be regulated and respond to the requirements of the cell. Several mechanisms of control in eukaryotes are also discussed, but these are not so well understood at the present time.

Questions

1 (a) Explain what you understand by the term 'gene regulation'.
  (b) Why do we assume that gene regulation should be more complex in multicellular eukaryotes than in prokaryotes?

(c) What will happen to the expression of genes in the *lac* operon of *E. coli* when a cell growing in the presence of lactose has a nucleotide pair deletion near the beginning of the z gene?

2 (a) Cite some evidence to support the idea that the activity of a gene is limited to certain stages during development.

(b) Explain the difference between gene regulation at the level of (i) transcription and (ii) translation.

3 What do you understand by the terms (a) gene amplification, (b) heterochromatinisation, (c) chromosome puffing?

18
Reproductive and breeding systems

In our study of genetics so far we have been largely concerned with the transmission, the structure and the action of the genetic material at the level of the cell and the individual organism. We must now turn our attention to another tier of complexity and consider how the process of heredity and variation work at the level of the **population** and of the **species** (Chapters 18–21). We will begin with the idea of the 'genetic system'. The term **genetic system** was first introduced by Cyril Darlington and it refers to *the way in which the properties of heredity and variation, methods of reproduction, and the control of breeding, are all bound up together in a group of organisms* (population or species). We know by now that heredity can be simply accounted for by the way in which DNA molecules replicate themselves and are then distributed into somatic cells and into gametes by the division cycles of the chromosomes in mitosis and meiosis. We know too that variation depends upon mutation and upon the processes of recombination which are associated with meiosis. What we must also appreciate is the way in which heritable variation is subject to the different kinds of reproductive and breeding methods that we find in various species.

The genetic system and variation

Three factors contribute to the control of genetic variation in natural populations: (1) mutation, (2) recombination and (3) the reproductive and breeding system.

1. Mutation   Mutation of genes and chromosomes is the basic source of all genetic variation. Gene mutations give rise to new alleles, and various forms of chromosome mutation allow for rearrangement of the genetic material and for the evolution of new genes and new types of chromosome (Chapters 15 and 16).

2. Recombination   Recombination takes place at meiosis by independent assortment of genes in different chromosomes and by crossing over between linked genes in the same chromosome. It is highly effective in producing variation among the offspring of heterozygous parents (Chapter 7).

3. Reproductive and breeding system   The control of variation depends upon the mode of reproduction, i.e. asexual or sexual, and upon the kind of sexual breeding, i.e. inbreeding or outbreeding.

Reproductive and breeding systems

Asexual reproduction

Asexual reproduction *is based on mitosis.* It involves only one parent and the offspring are all genetically identical, apart from any spontaneous mutations. *A population of cells or organisms derived by mitosis from a single cell or common ancestor* is known as a **clone**. Asexual reproduction is common among plants and among unicellular organisms such as the protozoa and one-celled algae. Only a few species reproduce exclusively by asexual means. *Amoeba* is one of them; it multiplies by binary fission.

There are numerous ways in which asexual reproduction can occur. They include means such as the production of asexual spores in fungi, budding in *Hydra*, twinning in man (separation of the zygote into two cells to give identical monozygotic twins) and various forms of vegetative propagation in plants. The term **vegetative propagation** usually refers to any way in which *part of the plant becomes detached to form a new independent offspring* (cuttings taken by man, runners, stolons, suckers, rhizomes, tubers, corms, bulbs, etc.). Two methods of asexual reproduction which are not readily identified as such, namely **parthenogenesis** and **apomixis** are dealt with in more detail in Box 18.1.

Box 18.1   Parthenogenesis and apomixis

Parthenogenesis is *the production of an embryo from an unfertilised egg.* The term is generally used with reference to animals, but strictly speaking it may be applied to plants as well. What happens in most cases is that the egg cell is formed by mitosis, instead of meiosis, and therefore contains an unreduced chromosome number. It develops without fertilisation and in animals all the progeny are diploid females. Some species have **facultative parthenogenesis**, which means that they have the capability of reproducing by this means as well as by the normal sexual process. Aphids are an example: in the summer months wingless females multiply up in this way by producing large numbers of genetically-identical wingless daughters. In other cases parthenogenesis is **obligate** and all members of the species are females. This happens in some roundworms, earthworms, sawflies, moths and the lizard (*Lacerta saxicola*). In the hymenopteran insects (e.g. wasps and bees) parthenogenesis is used as a means of regulating the sex ratio. A queen bee lays either fertilised eggs, which become diploid females, or unfertilised ones (formed as a result of meiosis in this case) which develop into haploid males. Parthenogenesis serves an important function in animals which are polyploid. It bypasses the complications in chromosome pairing and separation at meiosis, which often leads to inviability of the gametes (Chapter 16).

Apomixis occurs only in plants and is generally taken to mean *seed production without fertilisation.* It can happen as a result of parthenogenesis, by development of an 'unreduced' egg cell, or by development of the embryo sac from some other non-reproductive diploid cell of the ovule. The seeds resulting from apomixis are genetically identical clones. It is a particularly important form of reproduction in species hybrids, and polyploids which are sexually sterile. Apomixis occurs widely in flowering plants (and in ferns): two families where it is widespread are the Rosaceae and Compositae. The dandelion (*Taraxacum officinale*) is an obligate apomict. Other examples are the hawkweed (*Hieracium*), spring cinquefoil (*Potentilla tabernaemontani*) and the bramble (*Rubus fructicosus*).

Asexual reproduction can serve as a means of rapidly increasing populations of genetically identical individuals when conditions are favourable for the species, and as a way of survival when conditions are unsuitable for sexual reproduction. In certain types of polyploid plants (especially triploids), asexual reproduction circumvents the sterility which arises due to inviability of their gametes (Chapter 16).

Sexual reproduction

Sexual reproduction, as explained in Chapter 3, is the regular alternation in the life cycle of meiosis and fertilisation which provides for the production of offspring. We think of sexual reproduction as a means of bringing together the chromosome complements of two different parents, by fertilisation, so that their genes may be shuffled up at meiosis and then distributed into the progeny in a variety of new combinations (Chapter 7). This indeed is what happens in the majority of species, but in others it does not. In some species sexual reproduction may involve only one parent (self-fertilisation in plants), or two parents which are closely related, and the chromosome sets which come together in the zygote are then of identical, or very similar, genotype. Meiosis takes place in the same way, but it is ineffective in causing recombination because there is homozygosity at most gene loci. The **breeding system**, by which we mean whether a species is 'inbreeding' or 'outbreeding', is therefore an important factor in the control of genetic variability in natural populations.

Inbreeding is *any deviation from random mating*, i.e. towards mating involving *related* individuals. The most extreme form is self-fertilisation. An important genetic consequence of inbreeding is that it leads to homozygosity. We can easily see how this comes about by following what happens to the progeny when an individual which is heterozygous at a single locus (**Aa**) is repeatedly selfed (Fig. 18.1). The outcome is that the population quickly becomes homozygous at that locus and produces two pure-breeding lines, **AA** and **aa**. In reality,

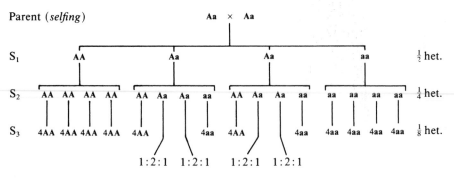

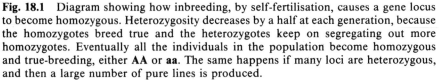

Fig. 18.1   Diagram showing how inbreeding, by self-fertilisation, causes a gene locus to become homozygous. Heterozygosity decreases by a half at each generation, because the homozygotes breed true and the heterozygotes keep on segregating out more homozygotes. Eventually all the individuals in the population become homozygous and true-breeding, either **AA** or **aa**. The same happens if many loci are heterozygous, and then a large number of pure lines is produced.

of course, there are many gene loci present in the eukaryotic genome, and therefore many different ways in which lines may become homozygous e.g. **AAbbccDDeeff** ... etc., **AABBccddEEff** ... etc., **aaBBCCDDeeFF** ... etc., and so on. Natural populations of inbreeders may therefore be composed of a number of different pure lines which all breed true.

There are various degrees of inbreeding, in different species, which are not as extreme as that of self-fertilisation. These include 'sib mating' (i.e. brother × sister), parent × offspring and mating between cousins. In all these cases the genetic consequence is much the same, but the time taken to attain homozygosity is usually much longer than with selfing. Mating between close relatives is the way that animal breeders obtain pure lines of livestock, and fanciers produce their pure-bred pedigrees.

There are three important implications of being inbred and homozygous:

1. *Meiosis is ineffectual in generating variability.* Recombination is meaningless in an individual which is completely homozygous. The progeny are genetically uniform and identical to the parent from which they derive (Fig. 18.2).

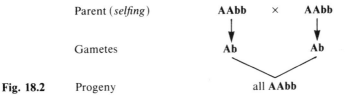

Fig. 18.2

2. *Variability is exposed.* Since there are no heterozygotes in the population, recessive alleles cannot be hidden in the heterozygous form. They are exposed as homozygotes and tested by natural selection immediately. Unfavourable ones are removed from the population.
3. *There is immediate fitness at the expense of long term flexibility.* In inbreeders all of the genetic variation that is present exists as differences between the various homozygous lines that make up the population. Because there is no crossing between lines this variation is 'fixed' and cannot be released. The lines breed true and each produces progeny which are identical to one another and to their parents. If the parents are well adapted to their environment then *all* of the progeny are also well adapted and there is immediate fitness. There is no wastage due to variable genotypes which are not well adapted. In the short term, therefore, inbreeding may be advantageous, particularly where conditions are of an extreme kind, or where a species is colonising such an isolated area that there may be problems in finding a mate. The disadvantage of this method of breeding is that it lacks long-term flexibility and makes no provision for the future. When the environment changes there is a restricted pool of variability from which to draw, and if none of the lines are viable in the new circumstances then they will all be doomed. Without the regular release of variation in each generation, inbreeders have no capacity to adapt, and in evolutionary terms they have no future—unless they can change to a measure of outcrossing.

Inbreeding can be provided for and be enforced in various ways. Some animals, such as the tapeworm, are hermaphrodite and lead such highly specialised lives that they have to fertilise themselves—they don't have access to a mate. As it happens though, most of the mechanisms which promote or enforce inbreeding are found in plants. There are three common ones based on the structure and development of the flowers. **Cleistogamy** is a mechanism in which the flowers fail to open: it is literally *fertilisation within closed flowers.* It is well known in *Viola.* A second system which is almost as effective, is where the flowers open but only after the pollen has been shed onto its own stigma (pea, *Pisum sativum*; and barley, *Hordeum vulgare*). In the third system the flower opens but the stigma is so enclosed within the stamens that self-pollination is inescapable (tomato, *Lycopersicum esculentum*).

Outbreeding is *random mating,* and its consequences are the opposite to those of inbreeding. It maintains heterozygosity and variation. Two parents which differ in their gene loci *combine* their alleles together into a heterozygote. The differences are then *recombined* at meiosis and distributed into the progeny as a variety of different genotypes and phenotypes in the next generation (Fig. 18.3).

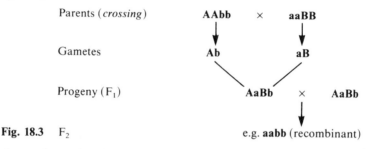

Fig. 18.3 F_2

As we have described in Chapter 7, this process of recombination is so effective that in natural populations of outbred species virtually every individual is genetically unique and distinct from every other. The offspring from any particular pair of parents differ from one another as well as from their parents. It follows that if the parents are well adapted, and suited to the conditions of their environment, then many of their offspring will not be. There is therefore some sacrifice of immediate fitness. If the environment should change, however, some of the variants generated by recombination may be better suited to the altered conditions than their parents. Indeed, if the change is drastic such individuals may be the only ones capable of surviving. The variability present in an outbred population is therefore important in terms of future adaptability and long-term flexibility. Populations of outbreeding organisms, because of their heterozygosity, generate a pool of variability in every generation. This variation can be drawn from immediately, should environmental conditions change and different combinations of genes become favoured by selection (see Chapter 20).

Genetic systems based on outbreeding are therefore more flexible than those based on inbreeding, and make more provision for the future. Outbreeders are more adaptable and their chances of survival in the long term are greater than those for inbreeders. The capacity for change also means that they are able to survive under a greater variety of conditions. In view of the greater advantages, and long term evolutionary security, it is not surprising that outbreeding occurs in the majority of existing species and a variety of mechanisms have evolved which prevent inbreeding.

These mechanisms include sexual differentiation into male and female forms, which applies to most species of animals (Chapter 8). Sexual differentiation precludes self-fertilisation, of course, but not other forms of inbreeding: these are generally avoided by complex patterns of behaviour. Outbreeding is also promoted by mechanisms such as mating types in fungi, and a whole range of pollination controls and systems of genetic incompatibility in plants.

Since we have already dealt with the genetic basis of sex determination in animals in some detail in Chapter 8, we will confine our remarks in the last part of this chapter to mechanisms which provide for outbreeding, and for genetic variation, in flowering plants.

Mechanisms of outbreeding in flowering plants

There are a number of species of flowering plants that have separate staminate ('male') and pistillate ('female') flowers. In some of them the two kinds of flowers are carried on the same plant, and in others they are borne on different plants.

Monoecious plants have both staminate and pistillate flowers on the same individual (e.g. hazel, *Corylus avellana*; and oak, *Quercus robur*). The tendency to cross-pollination (i.e. pollination between different plants) is often further enhanced by a timing difference in development of the male and female flowers on the same plant.

Dioecious species have staminate and pistillate flowers on different plants (e.g. poplar, *Populus alba*; and species of willow, *Salix*, Fig. 18.4).

(a)   *(b)*

Fig. 18.4 Flowers of the dioecious willow (*Salix*). (*a*) Catkin from a pistillate (seed-bearing) plant; (*b*) catkin from a staminate (pollen-producing) plant.

The majority of flowering plants are **hermaphrodite** and possess both stamens and pistils within the same flower. These types of flowers have various means which favour outbreeding and reduce the level of self-pollination and self-fertilisation to a minimum.

1. *Protandry.* In **protandrous** flowers the pollen is shed before the stigma is ready to receive it, and this naturally encourages cross-pollination (e.g. Canterbury bell, *Campanula*; and rose-bay willow-herb, *Chamaenerion angustifolium*). Self-fertilisation is still possible, however, by the transfer of pollen between flowers of different ages on the *same* plant.
2. *Protogyny.* **Protogynous** flowers have stigmas which are receptive before their pollen is shed (e.g. plantain, *Plantago major*; and figwort, *Scrophularia nodosa*). Self-fertilisation is again possible following pollination between flowers on the *same* plant.
3. *Incompatibility mechanisms.* These are physiological mechanisms whereby a stigma can distinguish between pollen of the same plant and that of a different plant from the same species. Such mechanisms are common in some groups of plants, such as the grasses and clovers, and they are highly effective in preventing self-fertilisation, even between different flowers on the same plant. Discrimination is based on recognition between proteins produced on the surfaces of the stigmas and pollen grains. Stigmas and pollen grains of the same plant carry identical proteins and are **incompatible**, i.e. the pollen fails to germinate or else the pollen tubes grow too slowly down the styles. Pollen from one plant is compatible with the stigmas of certain other plants because their surface proteins are different. Different forms of the proteins are determined by the alleles of a gene at the incompatibility locus. The mechanism is explained further in Fig. 18.5.

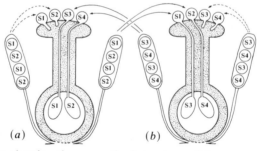

Fig. 18.5 Diagram showing the generalised structure of the sex organs of a flower and how the incompatibility mechanism works to prevent self-fertilisation. Proteins produced on the surface of the stigmas and pollen grains are determined by a gene which has many alleles (S_1, S_2, S_3, S_4 ... S_{100}). In an individual flower (*a*) the cells of the style and stigma are diploid and each cell contains one pair of alleles, say S_1S_2. Two kinds of proteins will therefore be secreted onto the stigma surface, since the alleles are co-dominant and both produce their protein product. Pollen grains of the same plant (*a*) will be either S_1 or S_2 (since they are haploid following meiosis), and will carry the same surface proteins, either S_1 or S_2. The recognition system is such that identical proteins give an incompatible reaction and selfing is prevented, i.e. the pollen fails to germinate or else the pollen tube grows too slowly. Cross-fertilisation can occur between plants (*a* × *b*) which carry different alleles because their pollen grains and stigmas produce different forms of the proteins which give a *compatible* reaction.

4. *Heterostyly.* The term **heterostyly** was first used by Charles Darwin in 1877. It describes *the presence of two or more different types of plant which are distinguished by the position of their stigmas and anthers within the flowers.* The best known example of pollination control by a mechanism based on such differences in floral morphology is in the primrose (*Primula vulgaris*). This example is dealt with in detail in Box 18.2.

Box 18.2   Heterostyly in the primrose

In the common primrose (*Primula vulgaris*) there are different types of flowers which are present on separate plants. The two main kinds of floral morphology found in natural populations are known as **pin** and **thrum**. The difference between them is mainly in the length of the style and the position of the anthers. In the pin type the style is long and the anthers placed low down in the corolla tube, whereas in the thrum type the style is short and the anthers are high up in the corolla (Fig. 18.6). Darwin recognised this heterostyly as a mechanism to promote outbreeding. Insects which visit these flowers, and act as pollinators, have a long proboscis which they extend down into the corolla tube in search of nectar at its base. When these insects visit a pin plant and extend their proboscis, pollen is deposited halfway along the proboscis length. This position is exactly appropriate for transferring pollen to the stigma of a thrum plant, but too low for depositing it on the pin stigma. The pollen therefore remains on the proboscis until the insect visits a thrum flower. Likewise when insects visit thrum flowers first, they will collect pollen high up on the proboscis. This is in the right place for dusting onto pin stigmas when the pollinator moves onto a pin plant, but too high for self-pollinating the thrum stigma.

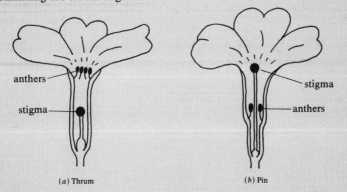

(a) Thrum   (b) Pin

Fig. 18.6   Difference in morphology between (*a*) thrum-eyed and (*b*) pin-eyed flowers in *Primula vulgaris.*

Heterostyly in the primrose thus favours cross-fertilisation between the pin and thrum-flowered plants, but it is only partly effective. When the proboscis is withdrawn from a pin corolla it takes up some pollen to its own stigma, and some pollen is likewise taken down from the anther of the thrum flower onto its stigma when the proboscis is inserted. Furthermore, in the thrum flower pollen from the ripe anthers just falls down onto the stigma. By far the most important pollination control mechanism in this species is not morphological at all, but is due to the *incompatibility* which exists between pollen and stigmas of the same flower—as we might expect, this incompatibility is stronger in the thrum flowers where the anthers are *above* the stigma.

The development of the pin and thrum forms is under the control of two major gene loci (at least we may look upon it in this simplified way). One of the genes determines the form of the style, so that **G** = dominant allele for short thrum type and **g** = recessive allele for long pin type. The other gene controls the position of the anthers, so that **A** = dominant thrum form with anthers high up the corolla and **a** = recessive pin type with low anthers. These two genes also control the physiological characters to do with the incompatibility reaction, as well as differences in certain other characters such as form of the stigma and pollen grain size. The two genes are tightly linked in the same chromosome as a 'supergene', giving the genotypes shown in Fig. 18.7.

Fig. 18.7

pin $\dfrac{g\ a}{g\ a}$ thrum $\dfrac{G\ A}{g\ a}$

A **supergene** is *a group of linked genes that are inherited as a unit*, and this may occur simply because they are so close together in the chromosome that crossing over rarely takes place between them (or for other more complicated reasons). In this case the linked pairs **g a** and **G A** behave as alleles of a single gene and segregate from the heterozygous thrum parent to give two kinds of gametes in equal proportions. There is no selfing, of course, and in crosses only two kinds of progeny are produced, pin and thrum, in equal numbers (Fig. 18.8).

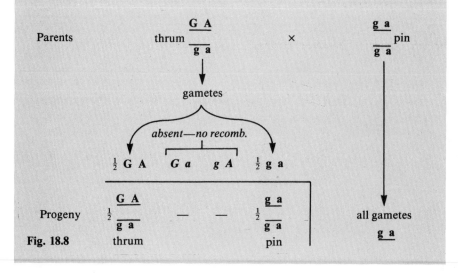

Fig. 18.8

Summary

In natural populations genetic variation is controlled by three components of the genetic system—mutation, recombination and methods of reproduction and breeding. This chapter deals with the influence of the reproductive and breeding system. Asexual reproduction and inbreeding are similar in that they both give rise to uniformity between parents and their offspring. Sexual outbreeding promotes genetic variation and most species have evolved mechanisms to ensure this kind of breeding system.

Questions

1 Fertilisation is not a universal phenomenon among higher animals since there are many animals which can reproduce parthenogenetically. However, relatively few reproduce exclusively in this way. In hymenopterous insects, such as ants, bees and wasps, fertilised eggs develop into diploid females but unfertilised eggs develop into haploid fertile males. The haploid males produce sperm by mitosis rather than meiosis. In other insects, such as aphids, the females produce only daughters during much of the breeding period. These daughters develop from unfertilised, diploid eggs.
   (*a*) Explain the meaning of the term 'parthenogenesis'.
   (*b*) Why do relatively few animals reproduce exclusively parthenogenetically?
   (*c*) What is the biological significance of:
       (i) haploid parthenogenesis to insects such as honey bees?
       (ii) diploid parthenogenesis to insects such as aphids?
   (*d*) What is the equivalent of diploid parthenogenesis in plants?   *(C. Biol., 1983)*
2 By reference to the life-cycles and reproductive processes of appropriate organisms, illustrate the ways in which sexual reproduction promotes the survival of the species in adverse or changing environmental conditions.   *(S.U.J.B. Biol., 1983)*
3 Describe, by reference to named examples, those mechanisms which favour cross-fertilisation in flowering plants.   *(O.L.E. Bot., 1983)*
4 What do you understand by the terms 'out-breeding' and 'in-breeding'? With reference to examples of both out-breeding and in-breeding species, discuss the relative advantages and disadvantages of these two systems.   *(O.L.E. Bot., 1981)*

19
Genes in populations

So far in this book, our studies on inheritance have been concerned only with experimental crosses. We have always started off with matings between two pure-breeding parents, and have then followed the progenies through the F_1 and F_2 generations, or else through various backcrosses.

At the level of the *population* the situation is different. In any particular generation *all* of the adults in the population are potential parents, and with sexual outbreeding they can interbreed in all kinds of combinations. For a gene with two alleles, **A** and **a**, there are three possible genotypes, **AA**, **Aa** and **aa**, and these may occur in any proportions relative to one another. This means that instead of dealing with well defined ratios within families we will be concerned with **frequencies** of the different genotypes in the population as a whole. The question that arises, therefore, is that given certain genotype frequencies, say 36% **AA**, 48% **Aa** and 16% **aa**, in one generation, what will be their relative frequencies in the next generation, and in the generation after that? In other words: how are genes inherited in populations?

In this chapter we will see that inheritance at the population level takes place by the same Mendelian processes of segregation and random combinations of pairs of alleles that we have studied earlier, and can easily be predicted by using a simple formula. *The study of the genetic composition of populations* is known as **population genetics**. The population geneticist tries to find out the frequencies of gene alleles, and genotypes, in natural populations and to study the factors which determine them. We will begin by defining and explaining some of the basic terms used in population genetics.

Populations, gene pools and gene frequencies

Populations

To the geneticist the term 'population' has a precise meaning. A **population** is *a local community of a sexually reproducing species in which the individuals share a common gene pool* (Fig. 19.1). Geneticists generally restrict themselves to populations of sexually-reproducing organisms in which there is **random mating** (**panmixis**) with *each member having an equal chance of mating with any other member of the population*. Because the sharing of genes takes place essentially by Mendelian inheritance, such local communities are also referred to as **Mendelian populations** (or **demes**).

The largest exclusive group of potentially interbreeding individuals which can comprise a Mendelian population is the **species**: but it is rare for an entire species to form one random mating group. What we generally find is that the species is made up of a large number of **local populations** (demes) with varying degrees of 'gene flow' between them. At one extreme we may have an 'open population' which is subject to immigration of genes from other intercommunicating groups within the species, and at the other end of the range a 'closed population' with the only source of new alleles being mutation. Generally speaking there are no clear-cut boundaries between one Mendelian population and another but, leaving this difficulty aside, the population geneticist looks upon the Mendelian population as the basic unit of study and is interested to know how the genes are distributed and inherited within these populations, and how one local interbreeding group differs from another in its genetic composition. The kind of population we have in mind might be a local cluster of wild garlic (*Allium ursinum*) growing in a wood; an isolated colony of the scarlet tiger moth (*Panaxia dominula*) in Southern England; a colony of house sparrows (*Passer domesticus*) inhabiting the buildings and hedgerows of an isolated village; a colony of black-headed gulls (*Larus ridibundus*) nesting in sand-dunes; or sticklebacks (*Gasterosteus aculeatus*) breeding in a pond. Throughout the rest of this chapter we will take it that we are concerned with sexual outbreeders, and we will not attempt to deal with any of the complications that arise from inbreeding and from various forms of **assortative**, i.e. non-random, mating.

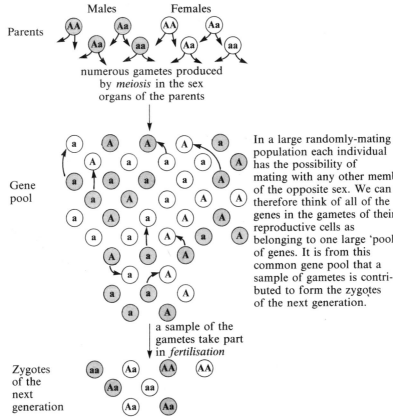

In a large randomly-mating population each individual has the possibility of mating with any other member of the opposite sex. We can therefore think of all of the genes in the gametes of their reproductive cells as belonging to one large 'pool' of genes. It is from this common gene pool that a sample of gametes is contributed to form the zygotes of the next generation.

Fig. 19.1 Diagram explaining the idea of a gene pool, using one gene with two alleles.

Gene pools

In a Mendelian population the sum total of the genes within the reproductive cells of all its members constitutes the **gene pool**. Reproduction in a population takes place, of course, between individuals of opposite sexes. But because mating is random, with each individual having an equal chance of mating with every other member of the population, we can think of all the genes in the gametes as belonging to one large 'pool' of genes (Fig. 19.1). It is from this pool that a **sample** is taken into the zygotes of the next generation. In thinking about the inheritance of genes in populations it is the gene pool, and the way in which its composition may or may not change over a number of generations, which must concern us—not what happens to any individual member. The unit of inheritance, and of evolutionary change, at the population level is the sample of the gene pool that is passed on from one generation to the next.

Gene frequencies

A gene with two alleles (**A** and **a**) will give rise in a population to three different genotypes (**AA, Aa, aa**); a gene with more than two alleles will give rise to correspondingly more. *The relative proportions of the various genotypes present in a population* form the **genotype frequency**.

The **gene frequency** (or **allele frequency**) is *the relative proportion of the alleles of a gene present in a population*.

The terms 'gene frequency' and 'allele frequency' have exactly the same meaning and they may be used interchangeably: modern textbooks favour the use of 'allele frequency', because it is the more correct term, but examination syllabuses generally use 'gene frequency'. Gene frequencies and genotype frequencies are important because they are characteristics of a population which the geneticist can estimate, and observe for change; and also because they describe the genetic composition of a population for a particular gene. In the study of population genetics we rarely deal with more than one or two gene loci at the same time—otherwise the system is too complex.

In order to distinguish clearly between the terms 'genotype frequency' and 'gene frequency', and to see how we estimate their values, we will use the **MN** blood group alleles in man as an example. The gene to which these alleles belong gives rise to the production of antigenic proteins on the surface of the red blood cells during their formation. There are two forms of the antigen, one determined by the **M** allele and the other by **N**. Homozygotes (**MM, NN**) carry the one form or the other and the heterozygotes (**MN**) produce both. The alleles are therefore *co-dominant* and each of the three genotypes can be easily distinguished and classified (see Chapter 10).

In a sample of 730 Australian Aborigines the numbers of individuals with the various blood group types were found to be as follows:

Genotype	MM	MN	NN
Number of individuals	22	220	488
Genotype frequencies (%)	3.0	30.1	66.9

The genotype frequency is the percentage of each genotype in the total sample, e.g. $22/730 \times 100 = 3.0\%$.

In the same way that we think of each individual as being representative of one genotype, so we think of each diploid individual within a population as being represented by one pair of alleles for the purpose of calculating gene frequencies (we have to overlook the fact that each individual is made up of many millions of cells).

To calculate gene frequencies, using numbers of genotypes, the procedure is as follows:

Genotype	**MM**	**MN**	**NN**		
Number of genotypes	22	220	488		
Number of **M** alleles	44 +	220			
Number of **N** alleles		220	+ 976		
Total number of alleles	44 +	440	+ 976	= 1460	

$$\text{Frequency of allele } \mathbf{M} = \frac{44+220}{1460} \times 100 = 18\%$$

$$\text{Frequency of allele } \mathbf{N} = \frac{976+220}{1460} \times 100 = 82\%$$

It is usual in population genetics to express both gene frequencies and genotype frequencies as decimal fractions, rather than percentages, so that the arithmetic can be handled more easily. The estimates of gene frequencies for the Australian Aborigines are therefore:

$$\text{Freq. } \mathbf{M} = 0.18$$

$$\text{Freq. } \mathbf{N} = 0.82$$

Since we are working with fractions, and we have only *two* alleles, then:

$$(\mathbf{M}+\mathbf{N}) = (0.18+0.82) = 1.0$$

To calculate gene frequencies using frequencies of genotypes, instead of numbers, the procedure is even simpler: it is based on the reasoning that all of the alleles in **MM** genotypes and half those in **MN** are **M**, and the same for **N**:

Genotype	**MM**	**MN**	**NN**	
Frequency (as decimal fractions)	0.03	0.30	0.67	
Frequency of allele **M**	0.03	$+\frac{1}{2}0.30$		= 0.18
Frequency of allele **N**		$\frac{1}{2}0.30$	+ 0.67	= 0.82

$$\text{i.e. Freq. } \mathbf{M} = \mathbf{MM} + \tfrac{1}{2}\mathbf{MN} = 0.18$$
$$\text{Freq. } \mathbf{N} = \mathbf{NN} + \tfrac{1}{2}\mathbf{MN} = 0.82.$$

Constant gene frequencies

The way in which genes are inherited in populations is no different in principle from the way in which they are transmitted in simple experimental crosses. All we require, at the population level, to predict the outcome of random mating among a mixture of genotypes is a simple extension of Mendel's idea of the segregation and chance combination of pairs of alleles that we described in Chapter 4. Provided that we know the frequencies of the two alleles (**A** and **a**) in the population, and provided that the three genotypes (**AA, Aa, aa**) all have equal chances of contributing gametes, then we can say precisely what frequencies we will have of **AA, Aa** and **aa** in the following, and in all subsequent generations.

Suppose, for example, that we have two alleles of a gene, **A** and **a**, present in a sample of gametes in frequencies of $0.6 \mathbf{A} + 0.4 \mathbf{a} = 1.0$. At fertilisation, the random combinations of these gametes will give genotypes among the progeny as shown in Fig. 19.2.

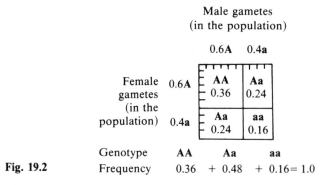

Male gametes
(in the population)

		0.6A	0.4a
Female gametes (in the population)	0.6A	**AA** 0.36	**Aa** 0.24
	0.4a	**Aa** 0.24	**aa** 0.16

Fig. 19.2

Genotype	**AA**	**Aa**	**aa**
Frequency	0.36	+ 0.48	+ 0.16 = 1.0

These genotypes will become the parents of the next generation and they in turn will contribute their gametes in frequencies of:

$$\mathbf{A} = \mathbf{AA} + \tfrac{1}{2}\mathbf{Aa} = 0.6 \qquad (\text{i.e. } 0.36 + 0.24)$$

$$\mathbf{a} = \mathbf{aa} + \tfrac{1}{2}\mathbf{Aa} = 0.4 \qquad (\text{i.e. } 0.16 + 0.24)$$

The gene frequencies in the gametes have therefore not changed between one generation and the next. The genotype frequencies will likewise be the same as they were before, and if we repeat the cycle over again we will find that both gene and genotype frequencies remain the same. In fact, the gene and genotype frequencies will remain constant *generation after generation*, indefinitely; provided that there are no disturbing influences (as described in Chapter 20) to prevent the different genotypes from making equal contributions to the progeny.

The reason for constant gene and genotype frequencies is the *binary* nature of inheritance. At fertilisation, alleles combine randomly in pairs in the diploid zygotes, and when they are present in frequencies of $0.6 \mathbf{A} + 0.4 \mathbf{a}$ they will always give genotype frequencies of $0.36 \mathbf{AA} + 0.48 \mathbf{Aa} + 0.16 \mathbf{aa}$ simply because these are the chance combinations for alleles in these frequencies. During meiosis, the opposite process takes place and segregation, or separation, of the pairs of alleles *undoes* these chance combinations and releases the alleles back into the gametes in their original proportions (Fig. 19.3).

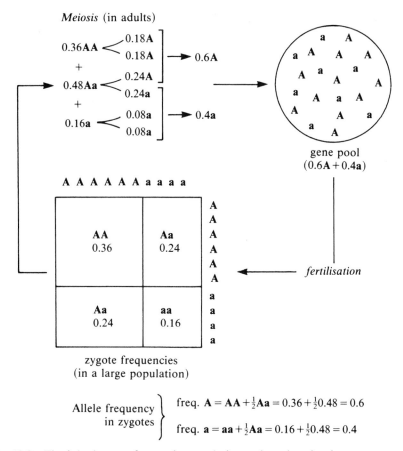

Allele frequency
in zygotes

freq. $\mathbf{A} = \mathbf{AA} + \frac{1}{2}\mathbf{Aa} = 0.36 + \frac{1}{2}0.48 = 0.6$

freq. $\mathbf{a} = \mathbf{aa} + \frac{1}{2}\mathbf{Aa} = 0.16 + \frac{1}{2}0.48 = 0.4$

Fig. 19.3 The inheritance of genes in populations takes place by the same processes of heredity as in simple Mendelian crosses. These processes involve the segregation, or separation, of pairs of alleles at meiosis, followed by their random combinations in pairs in the zygotes during fertilisation. In a *large* randomly-mating population the zygotes contain random (or chance) combinations of alleles, in frequencies which are representative of those in the gene pool; and when meiosis takes place these pairs of alleles are 'undone' and released back into the gene pool of the next generation in the same frequencies. Providing there are no disturbing influences (such as selection affecting some genotypes differently from others) this process of chance combination and separation goes on at each generation and the allele frequencies remain constant.

The Hardy–Weinberg law

The English mathematician G. H. Hardy and the German geneticist W. Weinberg discovered this rule of constant gene and genotype frequencies independently of one another in 1908. They showed that it could be used as a general law whatever the gene frequencies.

To see how this law works we will use the standard convention where p represents any given frequency of allele $\mathbf{A}$ and q (i.e. $1-p$) the corresponding value of $\mathbf{a}$. Gene frequencies in a sample of gametes are $p\mathbf{A} + q\mathbf{a} = 1.0$. In our example given above $p\mathbf{A}$ was 0.6 and $q\mathbf{a}$ was 0.4. The random combinations of these gametes give the genotype frequencies shown in Fig. 19.4.

Male gametes

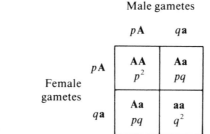

Female gametes

Fig. 19.4

This combining process may be represented more conveniently in the binomial expression:

$$(p\mathbf{A}+q\mathbf{a})(p\mathbf{A}+q\mathbf{a}) = p^2\mathbf{AA}+2pq\,\mathbf{Aa}+q^2\mathbf{aa}$$

or simply, $(p+q)^2 = p^2+2pq+q^2$

These genotypes become the parents of the next generations and we can calculate the gene frequencies in their gametes, in terms of p and q, as follows:

Freq. allele $\mathbf{A} = \mathbf{AA}+\tfrac{1}{2}\mathbf{Aa}$

$$= p^2 + \frac{2pq}{2}$$
$$= p^2 + p(1-p), \text{ because } q = 1-p,$$
$$= p^2 + p - p^2$$
$$= p$$

Freq. allele $\mathbf{a} = \mathbf{aa}+\tfrac{1}{2}\mathbf{Aa}$

$$= q^2 + \frac{2pq}{2}$$
$$= q^2 + q(1-q)$$
$$= q^2 + q - q^2$$
$$= q$$

The gene frequencies of $\mathbf{A} = p$ and $\mathbf{a} = q$ are therefore constant between one generation and the next; and the genotype frequencies will also remain unchanged, for the reasons given earlier (p. 258).

This relationship between gene and genotype frequencies is known as the **Hardy–Weinberg Law** (or **Principle**) and it is the basic rule which forms the foundation of all population genetics. The **Hardy–Weinberg Law** states that: *in a large randomly-mating population there is a fixed relationship between gene and genotype frequencies and, in the absence of selection, migration and mutation, these frequencies remain constant from generation to generation.*

The equation $(p+q)^2 = p^2+2pq+q^2$, which gives the relationship between gene and genotype frequencies, is known as the **Hardy–Weinberg Formula**.

Given the qualifying conditions mentioned above, we will expect the three genotypes **AA**, **Aa** and **aa** to be present in the same equilibrium proportions of $p^2 : 2pq : q^2$ in every generation, for whatever values of p and q we are dealing with, and to remain in these fixed frequencies for as long as the conditions needed for stability prevail. Under these circumstances a **genetic equilibrium** will exist and *the gene and genotype frequencies will be in accordance with the Hardy–Weinberg Formula.*

It also follows from what has been said above that only a single generation of random mating is required to establish a genetic equilibrium between gene and genotype frequencies. This is because genotype frequencies are determined only by the gene frequencies of the gametes which give rise to them. Any sample of gametes, with any given gene frequencies—irrespective of where they come from—will always give progeny genotypes in the proportions $p^2 : 2pq : q^2$, because these are the chance frequencies coming from randomly combining pairs.

In Chapter 20 we will discuss how gene frequencies can change, and how the *equilibrium* between gene and genotype frequencies can be influenced by the forces of mutation, migration, genetic drift and natural selection.

Applications of the Hardy–Weinberg formula

The Hardy–Weinberg Formula can be used by the population geneticist to study the genetic composition of populations and to test genotype frequencies to see whether or not they conform with equilibrium distributions. This can be done in situations where all three genotypic classes can be identified, as in the **MN** blood group genotypes. The gene frequencies are first found (as described earlier) and are used to calculate the expected distributions according to the Hardy–Weinberg Formula; these expected values are then compared with the observed frequencies of the population sample using the chi-squared test. When genotype frequencies do not conform to equilibrium distributions we may suspect that some disturbing influences are at work, such as mutation, migration, genetic drift or natural selection (Chapter 20).

A second application of the law is in finding the gene frequencies, and distribution of genotypes, in cases of dominance where we are unable to distinguish between the homozygous dominants and the heterozygotes on the basis of phenotype. Such an example is provided by the gene in man which controls the ability to taste the chemical phenylthiocarbamide (PTC). Tasting ability is conferred by the dominant allele T, so that both the homozygous dominant class (**TT**) and the heterozygotes (**Tt**) can taste this bitter substance, and the non-tasters are the double recessives (**tt**). If a large class of 200 students is tested it may well happen that 130 (65%) are tasters and 70 (35%) non-tasters.

Phenotype	Tasters	Non-tasters
Genotype	TT + Tt	tt
Frequency	0.65	0.35

We will follow the convention of representing the frequency of the dominant allele (**T**) by p and the recessive allele (**t**) by q; and $p + q = 1.0$. Making the *assumption* that the population is at equilibrium, and that we have none of the disturbing influences referred to above, the genotypes will occur in the following proportions according to the Hardy–Weinberg formula:

TT	Tt	tt
p^2 +	$2pq$ +	q^2

We know the value of q^2 to be 0.35, and by taking the square root we obtain the value of $q = 0.59$. Since $p + q = 1.0$, $p = (1 - q) = 0.41$. Substituting these values into the formula we can compute the genotype frequencies:

	TT	Tt	tt
	0.41^2	$2(0.41 \times 0.59)$	0.59^2
	0.17	0.48	0.35
or,	17%	48%	35%

A third use of the law is in making 'models' in order to see how gene and genotype frequencies change in response to certain aspects of selection. The use of the Hardy–Weinberg Law can also be extended to cover sex linkage, and the more complex genetics involved in working with multiple alleles, with two or more loci at the same time, and with polyploidy.

Summary

The geneticist looks upon a population as a large group of interbreeding individuals sharing a common gene pool. The unit of inheritance is the sample of this gene pool which is passed on from one generation to the next. Population geneticists try to estimate gene and genotype frequencies in order to study the genetic composition of populations and to find out how this sample of genes is inherited. It turns out that inheritance at the population level takes place by the simple Mendelian processes of segregation and random combination of pairs of alleles and can easily be predicted. The basic rule which is used for these predictions, and which forms the foundation of all population genetics, is the Hardy–Weinberg Law. When genotypes are undisturbed (by mutations or selection), and can make equal contributions to their progeny, this law predicts constant gene frequencies.

Questions and problems

1 (a) Explain what is meant by the term 'population', in the context of genetics.
  (b) What is meant by the terms 'gene frequency' and 'genotype frequency'?
  (c) Explain why gene and genotype frequencies remain constant from generation to generation in a population which is in Hardy–Weinberg equilibrium.
  (d) Describe briefly, using examples, how the geneticist may use the Hardy–Weinberg Formula to study populations.

2 An example of an inherited difference in man is the ability to roll the tongue into a U shape. Some people find this easy to do (tongue-rollers). Others find it impossible to do (non-rollers). The difference depends on a simple Mendelian alternative, in which the allele for tongue-rolling (T) is dominant to that for non-rolling (t). The following information was obtained from a sample of pupils in a school.

Tongue-rollers	Non-rollers	Total
490	210	700

In your answers to the following questions you are advised to show all your working clearly.
  (a) (i) What is the frequency of the t allele?
      (ii) What is the frequency of the T allele?

(b) Assuming that the Hardy-Weinberg principle applies, calculate the proportions of the school population which have the genotypes **TT, Tt, tt.**

(c) Is there a relationship between the dominance or recessiveness of an allele and the frequency of the phenotypic characteristic it controls in a population. Explain your answer.   (C. Biol., 1981)

3 In a randomly breeding population of mice agouti coat (**A**) is dominant to non-agouti (**a**). In a sample 16% were found to have non-agouti coat.

(a) What are the frequencies of the agouti and non-agouti alleles in the population?

(b) What proportion of the population would be expected to be homozygous for **A** and what proportion heterozygous?

(c) If the population continues to breed randomly what is the distribution of genes in the next generation?

4 Thalassemia major is a severe anaemia, usually fatal in childhood and frequent in Mediterranean populations. Thalassemia minor is a very mild anaemia, often difficult to detect at all.

(a) Among people of Southern Italian or Sicilian ancestry now living in Rochester, New York, thalassemia major occurs in about one birth in 2400, and thalassemia minor in about one birth in 25. Extrapolating these frequencies to a population of 10 000, the distribution is approximately as shown below.

Thalassemia major	Thalassemia minor	Normal
Th Th	**Th +**	**+ +**
4	400	9596

Verify that the frequencies of the **Th** allele and its normal alternative in the population are about 0.02 and 0.98 respectively.

(b) Does the population approximate to the binomial distribution of genotypes expected from this gene frequency?

5 Wild white clover (*Trifolium repens*) may contain chemicals which produce hydrocyanic acid (HCN) by hydrolysis when the leaves or stems are injured. Two dominant genes which assort independently are necessary for HCN release; gene **A** determines the production of the chemicals and gene **B** determines the production of the hydrolysing enzyme.

(a) Calculate the ratio of HCN-producing phenotypes to non-HCN-producing phenotypes in the F_2 generation when homozygous dominant and homozygous recessive parents are crossed, and the F_1 progeny are crossed with each other. (Show your workings.)

(b) The percentage of different phenotypes present in 800 plants collected in Bangor, North Wales was as follows:

AB—90%; **Ab**—6%; **aB**—3%; **ab**—1%.

Using the Hardy-Weinberg formula, calculate the probability of occurrence of the heterozygous genotype **Aa**. What is the expected number of these heterozygotes in the collection made at Bangor? Show how you arrive at your answer.   (W.J.E.C. Biol., 1982)

6 The blood cells of most human beings carry an antigen called the rhesus factor (**Rh**), and people with this antigen are described as rhesus-positive. People whose blood cells lack this antigen are described as rhesus-negative. Possession of this antigen is determined by a single gene.

(a) A rhesus-negative man married a rhesus-positive woman and they had two children. One child was rhesus-positive, the other rhesus-negative. Show diagramatically how this came about.

(*b*) When the children reached adulthood they each married a rhesus-negative partner. What is the probability of each of the following rhesus genotypes occurring in the first grandchild of the original parents?

Genotypes	Probability
Rh^+Rh^+	
Rh^+Rh^-	
Rh^-Rh^-	

(*c*) Blood samples were taken from 700 people at random, 112 of which were found to be Rh^-. Use the Hardy–Weinberg equation to calculate the number of individuals likely to be heterozygous for the gene. Show how you arrive at your answer.   (*W.J.E.C. Biol., 1980*)

7 (*a*) Explain the meaning of the term 'allele frequency' (often referred to as 'gene frequency').

(*b*) List three forces which may alter the allele frequency in a small population.

(*c*) (i) The algebraic expression of the Hardy–Weinberg Principle is given below:

$$p^2 + 2pq + q^2 = 1$$

where *p* and *q* are the frequencies of two alleles. State, in words, the Hardy–Weinberg Principle.

(ii) 'Woolly hair' is common among Norwegian families: the hair is tightly kinked and very brittle. The allele for woolly hair (**H**) is dominant over that for normal hair (**h**). The alleles for **H** and **h** have frequencies *p* and *q* respectively. In a certain population of 1200 people, 1092 individuals had woolly hair.

Assuming that the Hardy–Weinberg Principle applied, calculate the frequency of occurrence of each of the genotypes **HH**, **Hh**, and **hh**.

Show all your workings clearly.

(*d*) A short-fingered man with normal vision married a normal-fingered woman who was red–green colour-blind. The allele for short fingers is an autosomal dominant while the allele for red–green colour-blindness is an X-linked recessive. List the possible phenotypes of their sons and daughters.   (*A.E.B. Biol., 1983*)

20
Natural selection and speciation

The genetic variation that exists in natural populations provides the basis for change. Gene frequencies are not always constant in the way predicted by the Hardy–Weinberg Law (Chapter 19). Various forces act upon the variation and may alter the proportions of genotypes which are present in successive generations. *The gradual change in the genetic composition of a population over a number of generations* is known as **evolution** (organic evolution). In this chapter we will discuss the genetic basis of evolution and the way in which evolution can give rise to new species of living organisms.

Evolution

It could be argued that the theory of the evolution of species by natural selection, which was formally proposed in the Darwin–Wallace lecture to the Linnean Society in 1858, is the most important idea ever put forward in the field of biology. Until that time there were two conflicting schools of thought about where species came from. One belief was that all species were individually created and remain constant in their form. The other belief was that they could evolve, and that new species could gradually arise from pre-existing ones. This latter belief had been held by several eminent biologists including Erasmus Darwin, the grandfather of Charles Darwin, and the French biologist Lamarck (Box 20.1). The importance of the Darwin–Wallace lecture was that it offered an explanation of how evolution could take place—through the gradual process of change brought about by natural selection. This idea caught the imagination of the majority of biologists and is the basis of the theory of evolution which we accept today.

Box 20.1 Lamarck (1744–1829)

Lamarck subscribed to the view that species are not immutable, but could gradually change over long periods of time and evolve into new forms. According to his theory (1809) the modifications which an organism acquired during its lifetime, in response to the environment, could somehow be transmitted to its offspring. He believed that an organism's 'desire for improvement' was the driving force in evolution. The great size of the pectoral muscle of birds, for instance, he thought was due to the constant effort of straining to lift the bird into the air. Birds would improve their muscle structure, through the exercise of flying, and these improvements would then be passed on to their offspring. Likewise, organs for which there was little use would gradually diminish and become smaller in size and complexity. Lamarck collected many examples to support his theory and his idea made a great impact at the time. It was important because it involved the notion of 'evolution', and was quite at variance with the thinking of most of his colleagues who firmly believed that species arose by individual acts of creation and were immutable.

Lamarck's theory foundered because it was based on an idea of heredity which was wrong. He believed in the inheritance of acquired characters, "... all that has been acquired by their progenitors during their life is transmitted to new individuals ...". We now know that acquired modifications in the phenotype, as a result of the use or disuse of organs, cannot be inherited. Variations can only be transmitted if they are due to genetic differences. Lamarck had no knowledge of genetics and was unable to distinguish between genotype and phenotype, and thus between heritable and non-heritable variation.

Darwin and the theory of evolution by natural selection

Charles Darwin (1809–1882) devoted his whole life to the study of natural history and to a detailed series of observations and experiments in order to understand the adaptation of organisms, and their evolution. An important phase of his work began at the age of 22 when he accepted the position of naturalist on Captain Fitzroy's survey ship H.M.S. Beagle, which the British navy despatched on a five-year voyage around the world (1831–1836). On this voyage Darwin encountered the rich fossil beds of South America where he discovered many species of extinct animals and noticed the close resemblance in design between the fossil forms and the living species of, for instance, armadillos, tapirs and anteaters. He was greatly impressed too with the *variety* which he found within and between the different species of plants and animals. On the Galapagos Islands, off the coast of Ecuador, he noticed how different the principal groups of plants and animals were from those on the mainland. He observed too how the giant tortoises varied from island to island and distinctive *races* could easily be recognised by the form and pattern of their shells. Also important were the 13 species of finches which displayed remarkable adaptations suiting them to the different ecological niches which they occupied (p. 282).

In 1838, two years after returning from the voyage of H.M.S. Beagle, Darwin read an essay by the English clergyman Thomas Malthus who wrote about the growth of human populations. Malthus pointed out that unless population growth was checked in some way, by disease, war, famine or birth control, the number of people on the Earth would quickly increase until there was 'standing room' only. Darwin made calculations to show that the same geometric increase was true of any species. He had already begun to formulate his ideas about the origin of species by evolution when Malthus' essay gave him the idea of 'natural selection'. He could see that all organisms produce an excess of offspring and if they are all able to survive then their population would increase in size *exponentially* (2–4–8–16–32, etc.) In fact this does not happen, and numbers remain fairly constant from one generation to the next—except during the colonisation of new territory. He reasoned that the numbers are kept in check by competition for natural resources, such as space, food and mates. This competition he called the 'struggle for existence'. The variation that occurs between individuals of a species, such as minute differences in structure, behaviour and so on, affects their chances of survival and reproduction. He argued that individuals with variations which were advantageous, and which gave better adaptation to their environment, would be more likely to survive the struggle for existence, and to pass these favourable characteristics on to their offspring '...this preservation of favourable individual differences and variation, and the destruction of those that are injurious, I have called "Natural Selection".' Spencer's words, 'survival of the fittest', were later adopted for this part of the theory.

Darwin realised that in a given environment a species will gradually accumulate variations which are most suited to that environment. If circumstances change, a different set of variations become advantageous and will replace the previous less-well adapted forms. He knew also that to be selected the variations must be *heritable*—'...natural selection acts only by the preservation and accumulation of small inherited modifications...'.

Fig. 20.1   Charles Darwin (BBC Hulton Picture Library).

Darwin worked for many years on his theory of evolution, corresponding with other naturalists and amassing a wealth of data to support his ideas. To back his case for '...descent with modifications...' he drew on fossil evidence, comparative anatomy and embryology and on the great diversity of form shown within domesticated species of plants and animals (Chapter 21). Twenty years after reading Malthus' book he received a manuscript from another British naturalist, Alfred Russel Wallace, who had independently arrived at the same conclusion about the mechanism of evolution. Wallace graciously allowed papers by himself and Darwin to be read to the same meeting of the Linnean Society in 1858. The following year Darwin published his theory in a more extended form in his famous book *On the Origin of Species by Means of Natural Selection*. His work forms one of the corner-stones of biology.

Since Darwin's time a great deal more evidence has been accumulated to support the concept of organic evolution, which most biologists now accept as a reality. A brief summary of this evidence is given in Box 20.2. What is not so well understood, but remains a matter for much lively debate, is our understanding of *how* evolution takes place. Natural selection is generally accepted to be the main agent of change, but as we will explain later it is not the only one.

Box 20.2   Evidence for organic evolution

Fossil record

Fossils are *the organic traces, or impressions, of once-living organisms buried by natural processes and subsequently permanently preserved.* They show that living organisms did not all appear at once. The first forms were aquatic; terrestrial groups appeared much later. Some fossils show similarities with other fossils and also with certain living creatures, indicating a common ancestry. In general, the larger the age difference between a fossil and its living relative the greater the degree of divergence between them. This supports the idea of change by the gradual accumulation of many small differences over long periods of time. In some sedimentary rocks which are exposed at the edge of a cliff, or in a gorge, sequences of fossils can be found in strata of known ages, and the **palaeontologist** can show a series of gradual changes in the form of some organisms. Some evolutionary histories, such as that of the horse (*Equus*) have been reconstructed in great detail—the record suggests that it evolved from a small dog-sized creature (*Eohippus*, the dawn horse) with five toes on the fore legs and four on the hind, to the one-toed large animals of today. Certain link organisms have been found which have structures intermediate between recognised major groups. A well-known example is *Archaeopteryx*, a reptile-like animal with feathers, representing the transition from reptiles to birds.

Comparative anatomy of living organisms

Similarities in the basic structure of different groups of plants and animals indicate their evolutionary relatedness. Furthermore the comparison of **homologous structures**, *which are parts of an organism which have similar structure but different function,* supports the view that a basic structure can become modified or adapted to suit different needs. The best known example is the pentadactyl limb of vertebrates (Fig. 20.2).

They should not be confused with **analogous structures**, which are *parts with similar function but different structure,* e.g. a bird's wing and an insect's wing.

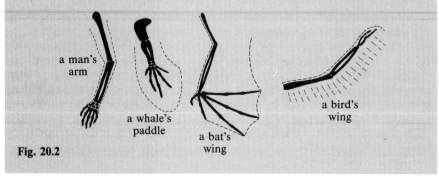

a man's arm

a whale's paddle

a bat's wing

a bird's wing

Fig. 20.2

Embryology

Resemblance between embryos is usually much greater than that between adult forms. Embryos of vertebrates are almost all identical in early development (e.g. they all form gill pouches) and evolutionists argue that this reflects their close affinity and descent from a common ancestral form.

Comparative biochemistry

All living organisms (except for some viruses) have DNA as their genetic material. All share the same universal genetic code and use their genetic information in the same way to code for proteins and RNAs which are virtually common throughout the plant and animal kingdoms. Many basic physiological processes, such as respiration, are the same in all species. The sequence of amino acids in certain proteins, such as cytochrome c (a respiratory pigment in all cells), reveals remarkable patterns of **molecular evolution**. The number of amino acids by which various species differ in their cytochrome c can be used to construct a tree of their evolutionary relationships. Man and the chimpanzee share an identical sequence of 112 amino acids. These sequencing studies suggest a common origin for the cytochrome c gene in all present day species.

Geographical distribution

The worldwide distribution pattern of animals, and of plants, can be related to the past history of continental land masses. The most distinctive faunas, for example, are found in Australia which is the most isolated continent: it separated from mainland Asia before the placental mammals appeared. The marsupials have been able to thrive there, presumably because of the absence of the more successful placentals.

Direct evidence

Selection imposed by man has brought about profound changes in the forms of our cultivated crops and domestic breeds of animals (Chapter 21). Changes in the genetic composition of populations, by natural selection (this chapter) and by human activities (Chapter 21), have demonstrated evolution in action.

Neo-Darwinism

Darwin's theory of evolution by natural selection had two major flaws: (1) he had no idea about the mechanism of heredity and (2) he was never able to demonstrate the process of natural selection actually taking place. He could not explain how variation arose or how it could be maintained over successive generations. This knowledge only became available after the rediscovery of Mendel's work and the development of the science of heredity. The advent of population genetics, which came much later, made it possible to analyse changes in the genetic composition of populations and to demonstrate the forces responsible for evolutionary change.

The re-examination of Darwin's theory in the light of discoveries about the physical basis of inheritance (genes and chromosomes) is known as **neo-Darwinism** (neo = new).

Sources of variation

Darwin believed in blending inheritance and in various other ideas about how characters were transmitted, which turned out to be quite erroneous. As it happened, his basic theory did not depend on knowing the mechanism of heredity, although he was criticised for being unable to explain the origin of his 'spontaneous variation'. We now know that the origin of genetic variation is due to the mutation of genes and chromosomes (Chapters 15 and 16). We know too that much of the variation we see in a population of outbreeders is due to recombination. Because of the particulate nature of heredity, alleles that combine together to form a heterozygote (**Aa**) retain their identity and can segregate out again unchanged (as **A** and **a**) in the gametes following meiosis. They can then *recombine* with the alleles of other genes to give an infinite variety of genotypes in the next generation (Chapter 7). This process of segregation and recombination can go on indefinitely as long as there is a supply of alleles available in the population.

We have seen earlier that all genes mutate at a constant and characteristic low frequency, about once per 100 000 gametes. The alleles which result are usually recessive and deleterious (Chapter 15). Because their frequency is so low they may persist for several generations before they are eventually exposed and eliminated as homozygotes, i.e. when two heterozygotes mate (**Aa** × **Aa**) some recessive homozygotes (**aa**) will appear among the progeny and will be selected against. *The elimination of segregating individuals which are homozygous for deleterious recessives, or which carry deleterious dominant alleles,* is known as **genetic death**. It may take the form of mortality (e.g. albino plants), sterility, failure to find a mate or any other circumstance which leads to reduced reproductive capacity. Natural populations are therefore genetically imperfect and they carry a burden, or load, of hidden genetic defects. In every generation a certain proportion of these recessives will segregate out as homozygotes and their presence will reduce the average level of *fitness* of the population. *The extent to which the burden of deleterious alleles causes the population to depart from its optimum fitness* is called its **genetic load.**

The genetic variability which is present in natural populations is the source of material upon which evolutionary forces act. Alleles and gene combinations which are unfavourable to an organism in one environment may well be advantageous in another one. Variation therefore gives a population the potential to change and to evolve in response to its environment. **Adaptation** is the term used to describe *the structural or functional features that have evolved in an organism and enable it to cope better with its environment.*

Natural selection: the principal force of change

Various forces operate upon the genetic variation present in natural populations to change their genetic composition and bring about adaptation and evolution. The principal force is natural selection, and this is the one to which we will give most of our attention.

Natural selection, in the way that we now see it, is defined as *a process that determines gene frequencies in populations through unequal rates of reproduction of different genotypes.* It comes about because the genotypes which comprise an outbreeding population vary in many aspects of their structure, physiology and behaviour, and some are better adapted to their environment than others.

Where there is some impairment to the functioning of an individual, in its particular environment, then it will leave less than the average number of offspring and will contribute correspondingly fewer of its genes to the next generation. The gene, or genes, which cause the impairment will be selected against and may even be eliminated from the population. Those which are lethal will obviously disappear much more quickly than those with less harmful effects, because the selection against them will be that much stronger. Equally, genes which improve the fitness of individuals will improve the organism's chances of survival and reproduction and will therefore increase in frequency. The fitness value of a particular genotype will depend upon the environment in which it finds itself, as we will see below.

There are thus three principles to natural selection that we need to bear in mind:

1. there is heritable variation between the individuals of a natural population;
2. these individuals have unequal chances of leaving progeny; and
3. these chances depend upon the conditions of the environment.

The best known examples of natural selection are those involving major gene **polymorphisms** where there are *two or more distinct forms of a species found in the same locality at the same time and in such frequencies (>1%) that the presence of the rarest form cannot be explained by recurrent mutation.* To begin to understand how selection works we have to study genetic variation of this kind which is controlled by only one or two major gene loci.

The peppered moth   One of the most straightforward and widely known cases of a change in gene frequencies by natural selection is that in the peppered moth, *Biston betularia*, in Britain. It is an excellent example of a **transient polymorphism** (i.e. the gradual replacement of one allele by another one).

The polymorphism involves a single gene locus with two main alleles. A recessive allele (**c**) determines the so-called 'typical' phenotype expressed in homozygotes (**cc**). These are light coloured with a 'peppering' of black spots on the wings and body (Fig. 20.3) and were the predominant form present in the first part of the 1800s. In 1849 a dark-coloured melanic variety called 'carbonaria' was reported for the first time in Manchester. The 'carbonaria' phenotype is due to a dominant mutation, **C** (**CC, Cc**). The two alleles are inherited in the predictable Mendelian manner in experimental laboratory matings, with **C** showing full dominance over **c**.

During the second part of the 19th century the dominant allele increased rapidly in frequency and by 1895 the 'carbonaria' variety comprised over 95% of the population in the Manchester area. The causes of this rapid alteration in gene frequency which Sheppard once referred to as 'the most spectacular evolutionary change ever witnessed and recorded by man', were extensively studied by Bernard Kettlewell in the 1950s. Kettlewell's work showed that the spread of the 'carbonaria' form was associated with the Industrial Revolution and the parallel spread of pollution caused by the fall-out from factory chimneys. The carbon particles and the SO_2 from atmospheric pollution killed the lichen on tree trunks and rocks and gradually changed the habitat of the moth (in the affected regions) from light-coloured lichen-covered surfaces to areas which were bare and blackened with soot. He showed a definite link between the pattern of spread of the 'carbonaria' and the presence of urban

(a) (b)

Fig. 20.3 'Typical' and 'carbonaria' forms of the peppered moth (*Biston betularia*) at rest on trunks. (a) Lichen-covered tree in the unpolluted Dorset countryside and (b) a lichen-free soot-covered trunk in industrial Birmingham. (From the experiments of Dr H. B. D. Kettlewell, University of Oxford. Photographs by John Haywood.)

industrial pollution. The highest frequencies of the melanic form were in the industrial regions themselves and also to the east of those areas where the pollutants drifted due to the effect of the prevailing westerly winds. This phenomenon of the spread of the melanic form of *B. betularia*, due to industrial pollution, became known as **industrial melanism**.

Kettlewell also demonstrated that the change in gene frequencies was due to natural selection and that the mechanism of selection involved differential predation by birds. This was shown by two kinds of experiment. In the mark–release–capture experiments, laboratory-reared moths were marked with spots of paint on the undersides of their wings and known numbers of the two forms were released in two contrasting woodland environments. One site was near to Birmingham city centre and was heavily polluted by soot (in the 1950s), the other was in rural Dorset and was free from pollution. When samples of the marked moths were recaptured a few days later the survival of the two forms could be compared. It was found that the dark 'carbonaria' form survived about twice as well as the typical form in Birmingham, but in Dorset the 'typicals' were at a much greater advantage. A second kind of experiment, using direct observation and cine-photography, confirmed that predatory birds were capturing moths which had been stuck on to trees. In the Birmingham wood the 'carbonaria' were better camouflaged on the darkened tree bark and were taken only about half as frequently as the 'typicals'. In Dorset the situation was the reverse, with typicals merging well into the background of the lichen-covered bark and the 'carbonaria' being much more exposed.

These field studies provided dramatic and convincing evidence of the 'force' of natural selection, in action, determining gene frequencies in natural populations. They gave an answer about what the mechanism was and made much of their impact from the demonstration that **micro-evolution** could occur quickly enough to be observed within the lifetime of a person, and does not necessarily require the centuries of time that Darwin thought were required. Kettlewell has referred to this work on the peppered moth as "Darwin's Missing Evidence".

During the past 30 years controls limiting the amount of atmospheric pollution by smoke have brought about a reversal of the environmental damage caused over the previous century, and the 'typical' form of *B. betularia* is now in the ascendant again and slowly increasing in frequency in Manchester and other industrial areas.

Sickle-cell anaemia in man   Sickle-cell anaemia is a genetically transmitted disease of the blood caused by an abnormal form of adult haemoglobin. We have already dealt with the structure of haemoglobin and the molecular basis of the defect which causes the disease (Chapter 14). The gene which controls this character is inherited as a single Mendelian recessive, with two alleles and three genotypic classes. Designation of the gene is **Hb** (for haemoglobin) and the alleles are Hb^A (normal haemoglobin A) and Hb^S (haemoglobin S). The three genotypes and their corresponding phenotypes are:

$Hb^A Hb^A$—normal homozygote

$Hb^A Hb^S$—heterozygous carrier with normal phenotype
(except under conditions of oxygen deficiency);

$Hb^S Hb^S$ —recessive homozygote with sickled cells.

Red blood cells of the afflicted homozygotes ($Hb^S Hb^S$) have a characteristic crescent or **sickle** shape (Fig. 20.4). They become sticky on their surface and clump together to cause interference with the blood circulation—they are rapidly destroyed, leading to anaemia, damage to the vital organs and eventually death. Four out of five sufferers die in childhood and many of the remainder soon afterwards. In the foetus, and in very young children, there is another form of haemoglobin (foetal haemoglobin, Chapter 17) which is not affected by the same mutation.

The surprising feature about this deleterious gene is that it occurs in very high frequencies (10–20%) in certain parts of the world—notably Tropical West Africa, some Mediterranean countries and parts of India—and the question is, why?

In the 1950s Tony Allison found the answer. He realised that sickle-cell anaemia is only prevalant in areas of the world where malaria is **endemic** (regularly found), and that there is a link between the two diseases. What happens is as follows:

1. The $Hb^S Hb^S$ homozygotes are at a disadvantage due to their high mortality from sickle-cell anaemia, and from a variety of other diseases to which they are also prone.

2. Homozygotes with normal haemoglobin ($Hb^A Hb^A$) are used as a source of food by the females of various species (about 30) of blood-sucking mosquitos (*Anopheles* spp.), which transmit the unicellular protozoan parasite

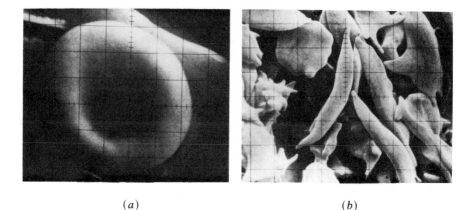

(a)   (b)

Fig. 20.4 Photomicrographs of red blood cells taken on a scanning electron micro-scope. (*a*) Single normal red cell at a magnification of ×10 000. (*b*) Sickle-cells ×5000. (The photographs were taken by Irene Piscopo of Philips Electronic Instruments on a Philips EM 300 Electron Microscope with Scanning Attachment. They are reproduced with permission from George W. Burns, *The Science of Genetics*, 5th edition, Collier-MacMillan, 1983. George Burns kindly supplied the photos.)

Plasmodium falciparum. The parasite is the real predator; it undergoes part of its life cycle within the red blood cells where it multiplies, bursts the cells, and gives rise to **malaria** fever.

3. Heterozygotes (**Hb^{A}Hb^S**) have a marked advantage over both classes of homozygotes because they are normally free of sickle-cell anaemia and they also have resistance to the malarial parasite. They have a mixture of both kinds of haemoglobin with sufficient normal molecules to enable the red cells to function properly, but their cells are an unsatisfactory source of nutrition for the parasite.

This heterozygous advantage leads to what is known as a **balanced polymorphism**. The heterozygote has the highest fitness level (i.e. the highest chance of survival) and therefore maintains both forms of the gene within the population, balancing out their loss through the mortality and reproduc-tive failure of the two homozygotes. The balance is in favour of the normal form of the gene (**Hb^A**) because malaria has a smaller effect upon the fitness of an individual than does sickle-cell anaemia.

In this example we can see how natural selection can act to maintain certain gene frequencies, and to preserve variation by keeping a deleterious gene within a population. If malaria is eradicated from a region, or if some members of the population move out of an area where it is endemic, then the heterozy-gotes lose their advantage. The polymorphism will then become *transient* and selection will gradually change the gene frequencies in favour of the normal allele.

The Snail   The snail, *Cepaea nemoralis*, has a highly-developed polymorphism which involves variation in the colours and banding patterns of its shell. A number of different genes are involved at separate, but quite closely linked loci. The two most important ones are those determining (1) the background colour of the shell and (2) the pattern of dark bands or stripes which encircle the shell.

1. The shell background colour gene has a series of multiple alleles specifying the colours brown, pink and yellow: each one in the series being dominant to the ones following.
2. Banding is controlled by two alleles which determine the presence of one to five bands, or their complete absence. Unbanded ($\mathbf{B}^0$) is dominant to any form of banding ($\mathbf{B}^B$) and when bands are present their number and pigmentation is influenced by genes at several other loci as well.

The various combinations of alleles of these different gene loci give rise in wild populations to a bewildering mixture of phenotypes, displaying a wide range of shell colours and banding patterns (Fig. 20.5). The frequencies, however, differ between one local habitat and another, and the differences appear to be consistent and well established. Studies on the banding of fossil shells shows that the polymorphism dates back tens of thousands of years and it seems certain that we are dealing here with what can only be described as a **stable polymorphism**. Not surprisingly, the question that population geneticists have been trying to answer for a long time is—what factors determine these local differences?

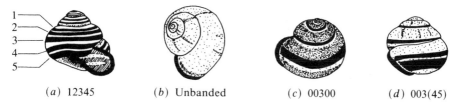

(a) 12345   (b) Unbanded   (c) 00300   (d) 003(45)

Fig. 20.5 Polymorphism for banding patterns and shell coloration in the snail *Cepaea nemoralis*. The drawings show a few of the more common varieties. Bands are numbered from the top of the largest whorl downwards. A missing band is denoted by 0 and fused bands are bracketed: (a) 5-banded yellow; (b) unbanded pink; (c) 1-banded brown; (d) pink shell with the first two bands missing and bands 4 and 5 fused. A shell with the upper two bands missing is said to be *effectively unbanded*, because viewed from above by a predator bands 4 and 5 are obscured by the curvature of the shell and band 3 is only partly visible.

Cain and Sheppard found out in the 1950s that natural selection on the basis of colour and banding phenotypes could play a major role in determining local patterns of the polymorphism. They studied populations (colonies) in the mixed environments comprising agricultural land and deciduous woods in the area around Oxford, and found a close matching of the colour patterns of the snails with the background in which they lived. Deciduous woods have a fairly uniform brown background of decaying leaves against which the effectively unbanded brown and pink shells (i.e. non-yellows) are relatively inconspicuous; whereas in hedgerows, grassland and green herbage the yellow-banded forms (which actually appear 'greenish' with the snail inside the shell) merge more readily into the vegetation. Differential predation by birds again provides the vital selective force which influences the composition of the populations. Song thrushes prey heavily on snails, particularly at nesting time, and they are selective in taking the ones which are most easily seen against their background. This is known because the thrushes break open the larger specimens on stones, or 'anvils', and the shell remnants provide a sample of what has been taken. Direct observations on the preferences for various forms, as well as the analysis of shell remains from paint-marked snails, have confirmed that the birds prey preferentially on the most conspicuous and least 'cryptic' forms.

It is evident, of course, that if all the selection were due to predators in this simple way then we would expect to find *only* yellow-banded shells in rough green herbage, and uniformly brown or pink unbanded ones in dense woodland—but this is not the case. Each of the local colonies sampled is polymorphic, and retains the rarer of the two main forms at an average frequency of about 20%. There must be another side to this story, as in sickle-cell anaemia, and some other agent or aspect of selection that maintains the polymorphism. In this case the nature of the balancing selection is not really understood, but there are three factors at least that may play some part in it: (1) frequency-dependent selection, (2) temporal changes in the environment and (3) heterozygous advantage.

1. **Frequency-dependent selection** means that the most obvious and unconcealed forms become more difficult to find as their frequency in the population declines. Their loss is then balanced by the predators switching temporarily to the more cryptic forms.
2. **Temporal change** refers to the way in which the background vegetation undergoes seasonal changes in its growth. This leads to some 'mismatching' at certain times of the year and to some heterogeneity of habitat which helps to maintain the polymorphism.
3. It is a well-known genetic phenomenon that heterozygous combinations of alleles often lead to an enhancement of the vigour and physiological fitness of an organism, relative to that of the homozygous condition. In the snails there are several different genes controlling the colour and banding patterns and many different ways in which heterozygotes can arise even in phenotypes which appear visually quite similar to one another. It seems highly likely, therefore, that selection could also be working *indirectly* to preserve the polymorphism by favouring heterozygotes on the basis of their superior physiological fitness (viability and fertility).

It is also known that there are differences in physiology which are associated with the different morphological forms. These relationships could be important in determining the distribution patterns of the snails in some other areas, where there is no matching with the background and where uniform phenotypes are present over large areas of land.

Heavy metal tolerance in grasses: selection for a quantitative character   In discussing natural selection so far we have confined ourselves to examples involving major gene variation. Selection also acts on quantitative characters, but this is much more difficult to study. The way in which we deal with it is described in Box 20.3.

We will consider just one example of natural selection acting upon continuous variation—that of heavy-metal tolerance in grasses.

Waste materials forming the spoil heaps of old derelict mine workings, where ores were once extracted, contain high concentrations of certain heavy

Box 20.3 Selection for quantitative characters

Genetic variation controlling quantitative characters is just as much affected by natural selection as that due to major gene polymorphisms, but the process cannot be described in the same way in terms of changes in gene frequency. With continuous variation the relation between the gene and the character is much less obvious: individual genes cannot be identified and the population geneticist has to monitor the progress of selection using means and variances (variation about the mean) of the population. Three main modes of selection for polygenes have been identified and described by Mather: they are illustrated in the graphs in Fig. 20.6.

Mode of selection

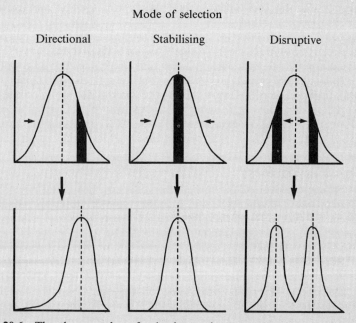

Directional   Stabilising   Disruptive

Fig. 20.6 The three modes of selection acting upon continuous phenotypic variation. Horizontal arrows indicate the direction in which selection is operating and the black areas show the new optimum phenotypes which will contribute to the next generation. Dashed lines represent the old and new means.

Directional selection operates in a changing environment and results in a reduction in variance and progressive shift in the population mean for the character concerned until a state of adaptation is reached. It is the main form of selection practised by man in the improvement of domesticated plants and animals.

Stabilising selection. The strongest selection pressure is against the extreme variants furthest from the mean; it leads to a reduction in variance without any change in mean. It is the form of selection which operates in a constant environment to maintain the best adapted genotypes within the population.

Disruptive selection favours two optimum phenotypic classes at the expense of intermediates. It occurs in natural populations where two distinct habitats, or different kinds of resources, exist, and in plant and animal breeding where selection is practised for the extremes of size and form.

metals such as lead, zinc, copper and nickel, which are toxic to plants and animals. These old mines are common throughóut Britain and the spoil heaps are often bare and devoid of vegetation even though some of them have been in existence for 100 years or more. The mines are a good hunting ground for ecological geneticists because they represent an extreme form of environment and on some of them a few species, particularly grasses of the genera *Agrostis* and *Festuca*, have evolved metal-tolerant races which are able to colonise these barren and contaminated sites. Breeding experiments have shown that the tolerance is genetically determined and is inherited as a quantitative character, showing many grades of tolerance. Grasses may become tolerant to several metals at the same time, when these are present together; but when this happens the tolerances to the individual metals are highly specific and genetically independent of one another.

In relation to natural selection one of the most interesting features of these metal-tolerant populations is how small an area they cover, and how they manage to maintain their adaptation in the face of a continual 'inflow of genes' from the non-tolerant populations which completely surround them.

The situation is exemplified very well by the work of McNeilly on the copper tolerance of *Agrostis tenuis* growing on a small mine in Caernarvonshire in North Wales. The situation is described in Fig. 20.7. To study the adaptation of the grasses McNeilly has made comparisons between the copper tolerance of the adult plants growing on the mine area with that of their progenies raised from seed. The findings can be summarised as follows:

1. Adult plants sampled on the mine, after selection has taken place, show a high mean index of tolerance and a small range of variation. Their seeds, grown in normal soil and therefore not subject to selection, show a mean index of tolerance which is much reduced and a wider range of variation. This indicates (*a*) that the genetic composition of the seed generation has been influenced by migration of genes (carried in pollen) into the mine area from upwind non-tolerant populations; and (*b*) that on the mine itself there must be strong directional selection imposed upon the seedlings at each generation in order to eliminate the low-tolerance phenotypes and maintain the adaptive characteristics of the adult population.
2. Seed produced by copper-tolerant plants which have been grown in isolation, under experimental greenhouse conditions, maintain a high mean level of tolerance but are much more variable than adult plants growing on the mine. Because the plants which produced these seeds were grown in isolation they were protected from the pollen of non-tolerant plants, and from an inflow of their genes. Their variation is greater because they are unselected and they display variability which results from the segregation and recombination of their polygenes.

Strong selection pressure for metal tolerance, at the seedling stage, is therefore responsible for maintaining the adaptation of the mine population and their sharp demarcation from adjacent non-tolerant plants. The overspill of tolerance genes into the surrounding downwind area occurs because selection against tolerant plants growing in normal soil is very much weaker. To some extent the gene flow from non-tolerant plants is also offset by reproductive strategies—plants in the mine populations flower earlier than those outside, and there is also a certain amount of self-pollination.

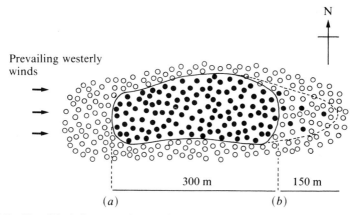

N

Prevailing westerly
winds

300 m   150 m

(a)   (b)

Fig. 20.7 Simplified diagram of McNeilly's copper mine in Caernarvonshire showing the distribution of tolerant (solid circles) and non-tolerant (open circles) plants. The continuous line represents the mine boundary. At the upwind end of the mine (a) the boundary between tolerant and non-tolerant populations is very sharp, while at the downwind edge (b) it is blurred and tolerant plants extend into the normal population for up to 150 m (area denoted by dashed line).

Other forces of evolutionary change

Gene mutations are rare events, occurring at the rate of about once in every 100 000 gametes for a given locus (Chapter 15). Mutational changes can alter allele frequencies in populations but the degree of change is thought to be quite negligible in comparison with other forces such as selection.

Migration (gene flow) is *the transfer of genes between populations by the movement of gametes, individuals, or groups of individuals from one population to another.* Its most important effect is to counteract local adaptation, and divergence, by introducing genes from one population into another one of different genetic make-up. In some cases, though, selection pressures can be so strong as to overcome the effects of migration, as in the copper-tolerant mine populations of *Agrostis.*

Genetic drift is *the random fluctuation in gene frequencies due to sampling error.* In a large population there is a strong likelihood that the sample of genes coming together in the zygotes will be fully representative of the gene pool in the previous generation. A small population on the other hand is much more prone to sampling error, and the genes taken into the zygotes may be quite *unrepresentative* of the gene pool due to chance alone. In a small population, therefore, the gene frequencies may fluctuate widely from one generation to the next.

The founder effect is a once-and-for-all change in gene frequencies that may take place when a small group of individuals wanders off, or becomes isolated, from the main population and 'founds' a new population. Because the group is small it may carry an unrepresentative sample of genes, due to sampling error, and the new population may differ widely in its genetic constitution from the original one. This can happen in the formation of island races, for example, or when the numbers of a population are suddenly reduced by some catastrophe. It is the most drastic way that is known of bringing about a sudden change in gene frequencies. Descendants of races formed in this way can differ from their ancestral population in many gene loci.

Speciation

There are almost as many definitions of the term 'species' as there are different species. Problems of definition arise because in taxonomic terms a species means organisms which share specific morphological characters, and they are often described from a few dead 'type specimens'. In nature, however, species vary in space and time, and there are often gradations in form leading to uncertainty about where one species ends and another one begins. The song sparrow (*Melospiza melodia*), for instance, is widely distributed throughout the United States and there are many local **races** which each have their own distinctive form and song pattern. Are they all members of the same species? The answer is yes, because, just as in man, wherever individuals from different geographical races come together they will interbreed and give rise to intermediate populations. It is this capacity for interbreeding which unites organisms into species and which separates one species from another. Races will interbreed but species will not. We can therefore define a **species** as *a group of interbreeding natural populations that share common morphological characteristics and are reproductively isolated from other such groups.* This definition obviously cannot apply to self-fertilising organisms, or to forms which reproduce solely by asexual methods. Nonetheless, such groups are frequently called species where they consist of individuals which are very similar to one another. The species is the largest group of organisms which share a gene pool. There are about three million such groups on earth at the present time, but their number and kinds are slowly changing all the time. Some of these groups are in a stable relationship with their environment, some are becoming extinct, and others are in the process of evolving into new reproductive groups—that is, undergoing **speciation**.

Mechanism of speciation

It is not possible to give a really clear-cut and definite account of the process by which new species arise. In some special cases, such as the 'instant speciation' which may result from polyploidy in plants (see below), the mechanism is well understood. But the majority of species do not evolve in such a simple or rapid way. The process is usually gradual and takes place over a long period of time—of the order of several thousands of years.

The general idea about speciation is that natural selection, and the other forces of evolutionary change mentioned above, may act differently in the different populations which comprise a species. When such differential action continues over a large number of generations the populations concerned may slowly diverge from one another, by the accumulation of numerous small genetic differences, and begin to form a number of different *races*. This process occurs because no two populations can occupy precisely the same niche, or exploit exactly the same kind of resources at the same time. As long as there is interbreeding between the populations, however, they will continue to exchange their genes and will remain united as one species. The critical event which allows the divergence to proceed beyond the level of races, and into new species, is the formation of some kind of *barrier* which prevents gene flow between populations. *Such a barrier which prevents the gene exchange is called an* **isolating mechanism**. There are a number of different isolating mechanisms, which are listed in Table 20.1, and therefore several different ways in which divergence and speciation may occur.

In discussing the mechanism of speciation it is customary to distinguish between that which occurs due to some method of physical separation of the gene pool into different geographic regions (**allopatric speciation**) and that which arises from an isolating mechanism within a gene pool in the same geographic region (**sympatric speciation**).

Table 20.1   A summary of the most important mechanisms which isolate closely related organisms into separate breeding groups.

1. **Geographical separation**

2. **Reproductive isolation**

   (*a*) **Prezygotic barriers** (prevent fertilisation and the formation of zygotes).
   - (i) *Seasonal:* populations exist in the same regions but are sexually mature at different times of the year, or in different years.
   - (ii) *Ecological:* populations live in the same geographic region but occupy separate habitats.
   - (iii) *Ethological* (in animals): isolation by species-specific mating behaviour.
   - (iv) *Incompatibility:* cross-fertilisation between species prevented by physiological incompatibility (e.g. pollen–stigma interactions, Chapter 18) or differences in structures of reproductive organs.

   (*b*) **Postzygotic barriers** (fertilisation can occur, but hybrids are either not formed or else they are sterile).
   - (i) *Hybrid inviability:* hybrids weak of developmentally abnormal.
   - (ii) *Hybrid sterility:* hybrid develops to maturity but is sterile due to failure of meiosis and abnormal segregation of chromosomes and genes.

We will confine ourselves here to discussing a relatively simple model of speciation—geographic speciation (Fig. 20.8).

1. The initial event involves some physical separation of an existing population into two or more sub-populations so that the free exchange of genes is prevented. This may occur, for example, when part of a land mass becomes detached as an island.
2. Once spatially separated, the sub-populations will diverge genetically. Their environments will be different, they will have different patterns of variation, and selection will act in different ways to alter their genetic composition.
3. Eventually this divergence may proceed to such an extent that the separate groups are no longer able to interbreed even if they are brought together again as sympatric populations. At this stage they are reproductively isolated and by definition have become different species.

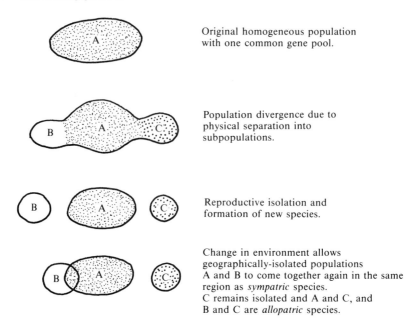

Original homogeneous population with one common gene pool.

Population divergence due to physical separation into subpopulations.

Reproductive isolation and formation of new species.

Change in environment allows geographically-isolated populations A and B to come together again in the same region as *sympatric* species. C remains isolated and A and C, and B and C are *allopatric* species.

Fig. 20.8   Diagram showing the sequence of events which lead to geographic speciation, starting with a homogeneous group of organisms (A) sharing a common gene pool.

Darwin's Finches: an example of geographic speciation

The Galapagos Islands were visited by Darwin in 1835 during the voyage of H.M.S. Beagle. They are a group of volcanic islands which were pushed up out of the sea more than a million years ago, and have never been connected to the mainland of South America. They lie 600 miles west of Ecuador. The land animals that inhabit them must have come across the sea from the mainland. Among the few species that have established themselves, there are a number of finches. Specimens of these birds were collected by Darwin and have subsequently been studied in great detail—notably by David Lack who spent a year on the Galapagos (1938–39). In all there are 14 species, 13 on the Galapagos and one on nearby Cocos Island: collectively the birds are known as **Darwin's Finches**.

Darwin was greatly influenced by the variety in these finches, particularly in the different sizes and shapes of their beaks which are associated with their different habits of feeding (Fig. 20.9). He realised that the islands must have been colonised by migrants from the mainland, where there is only one type of finch, and that their descendants had become diversified and adapted to the various ecological niches which they found. **Adaptive radiation** is the term now used to describe *the evolution of a group of organisms along several different lines involving adaptation to a variety of environments.*

It is envisaged that the different species evolved in isolation on separate islands, and some of them still remain isolated. Some of the others, however, now live as sympatric species on the same island. The different species co-exist

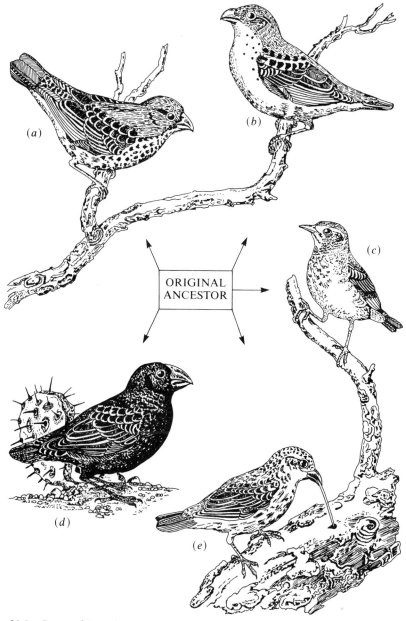

Fig. 20.9 Some of Darwin's Finches which originated on the Galapagos Islands from a common ancestor through adaptive radiation. They differ mainly in the form of their beaks and have different feeding habits which are associated with the different ecological niches available on the Islands. (*a*) Insect-eating tree finch. (*b*) Vegetarian tree finch with parrot-like beak for eating fruits and buds. (*c*) Warbler-like finch feeding on small insects. (*d*) Cactus-eating ground finch (the other ground finches are mainly seed eaters). (*e*) Woodpecker-like finch which climbs trees and uses a cactus spine to dislodge insects from crevices in the bark. When the insects emerge it drops the spine and seizes them with its beak. (After David Lack.)

in this way because they are reproductively isolated by mating behaviour, and there is no competition between them for food—they have evolved differences in diet. It is important to appreciate that this geographic speciation could take place on the Galapagos (and other similar situations) because the ancestors of the finches were able to colonise habitats that were largely unoccupied when they arrived. On the mainland this diversity of finches is not found—here the available niches were already filled by species of other birds. The woodpecker-like finch, for example, could never have evolved on the mainland in competition with the true woodpecker.

Polyploidy and speciation

Polyploidy is an important exception to geographic speciation, and is a means of bringing about reproductive separation in sympatric populations. It is much more important in plants than it is in animals, as discussed in Chapter 16.

When chromosome doubling takes place in a diploid, to give an autotetraploid, the newly formed polyploid will immediately be isolated from its parent species by a sterility barrier. The reason for this is that hybrids between the new $4x$ and the original $2x$ forms will be triploid ($3x$), and these are sterile due to irregular chromosome behaviour at meiosis (Chapter 16).

In allopolyploidy a new species can arise by hybridisation between two existing species, followed by spontaneous chromosome doubling (Chapter 16): again it will be reproductively isolated from both diploid ancestors due to triploid sterility. Cultivated wheat is an important example of the origin of a new species by allopolyploidy (Box 16.2, p. 227). Another well-known case is that of cord grass, *Spartina townsendii*. This species is abundant around the coasts of Britain where it has been widely used to stabilise mud flats. It is thought to have arisen about 1870 in Southampton water, by hybridisation between an introduced American species (*S. alteriflora*) and a native British species (*S. maritima*) to form a sterile hybrid *S. anglica*. Doubling in chromosome number gave rise to the fertile allotetraploid *S. townsendii*.

Summary

Species are not immutable. They undergo continuous gradual change and evolve into new forms. The principal agent of change is natural selection, which acts like a sieve and favours those forms which are best adapted to their changing environments. Genetics explains how evolutionary change takes place, and strengthens the theory of evolution by natural selection. Neo-Darwinism can explain the source of variation within a species, and can show how the genetic composition of natural populations can change over short periods of time due to micro-evolution. The study of population genetics has also revealed that there are forces of change other than natural selection: the principal one is genetic drift. Speciation is the result of natural selection acting in different ways upon homogeneous populations which become separated into different breeding groups. This separation comes about, and is maintained, by various methods of reproductive isolation which provide barriers to gene exchange.

Further reading

Berry, R. J. (1977), *Inheritance and Natural History*, Collins.

Bradshaw, A. D. and McNeilly, T. (1981), *Evolution and Pollution*, Studies in Biology No. 130, Edward Arnold.

Ford, E. B. (1973), *Evolution Studied by Observation and Experiment*, Oxford Biology Reader Series, Oxford University Press.

Sheppard, P. M. (1975), *Natural Selection and Heredity*, Hutchinson.

Shorrocks, B. (1978), *The Genesis of Diversity*, Hodder and Stoughton.

Stebbins, G. L. (1982), *Darwin to DNA, Molecules to Humanity*, Freeman.

Questions

1 State the theory of evolution put forward by Darwin and cite the evidence he used in formulating this theory. In what ways has the theory been modified by more recent evidence? *(L. Biol., 1980)*

2 Critically and impartially examine the evidence for and against the theory that living organisms, including animals, have evolved continuously on Earth.

   Equally impartially and critically examine the evidence that natural selection has been the main factor determining the mode of evolution. *(O.L.E. Zool., 1981)*

3 (*a*) Discuss what is meant by the term 'species'.

   (*b*) Explain why isolation is important in the origin of a new species.

   (*c*) Describe the different kinds of isolating mechanism. *(C. Biol., 1981)*

4 What is a species? Do species really exist?
(J.M.B. Nuffield Biol., Special Paper, 1982)

5 (*a*) Define the term 'gene pool'.

   (*b*) The gene pools of populations may be correlated with geographical distribution.

   (i) Geographical isolation can result in variation within animals without creating a new species. Name *one* example where this has occurred.

   (ii) Name *one* example which does illustrate the formation of new species as a result of geographical isolation.

   (*c*) If after many generations the separated groups are brought together again, reproductive isolation ensures that they remain distinct species. Suggest two ways in which reproductive isolation may be maintained.
(W.J.E.C. Biol., 1982)

6 Write an essay on mechanisms of speciation and species isolation.

7 What is meant by the term 'natural selection'? Comment on the importance of modern genetical studies in determining the mechanism of this process. Give examples which show that natural selection is occurring today.
(S.U.J.B. Biol., 1981)

8 Write briefly on *three* of the following:

   (*a*) response to selection;

   (*b*) industrial melanism;

   (*c*) genetic drift;

   (*d*) heavy metal tolerance in plants.

9 What is meant by 'genetic polymorphism'? Give *one* example of a transient polymorphism and one example of a stable polymorphism, and indicate the main factors that are of influence in your examples.

10 'Genes mutate, organisms are selected and groups evolve'. Discuss.
(A.E.B. Biol., 1983)

21
Selection imposed by man

In recent times man's activities have had a profound effect upon the population genetics of other species. Large areas of the Earth's land surface are now covered with domesticated forms of plants and animals. Wild populations are frequently subjected to various forms of chemical treatment aimed at their control, or else at their complete destruction and eradication. This applies particularly to species such as disease-causing organisms, which are directly harmful to humans, and to those which are indirectly competing with us for sources of nutrition—e.g. fungal pathogens of crop plants, weeds competing with crops for water, light and nutrients, animals which attack stored food products and predators of domestic animals.

In this chapter we will consider some aspects of selection imposed by man. The distinction between what we have to say here and some of the examples of selection given in the previous chapter is somewhat arbitrary. Obviously the selection which gave rise to the spread of melanic moths, and to copper-tolerant strains of *Agrostis tenuis*, were the result of man's activities. But they were not deliberate. Changes in the genetic composition of species and of populations included in this chapter are the result of deliberate interference with other organisms for our own gain and protection—although these changes are not necessarily all harmful. It is perfectly possible to domesticate and to breed from a wild species of plant or animal without unduly disturbing the natural populations from which it came.

Selection imposed by man is often more rapid and intense than that which occurs in nature. The selection pressures are very strong and can be applied to produce extremes of diversity and development. The results of these activities demonstrate the wealth of genetic variation that is available within a species and how it can be manipulated to produce large changes in the form and physiology of organisms. The variation that we see under domestication provides strong support for the theory of evolution by natural selection.

Domesticated plants and animals

All of the forms of plants and animals that we use today for our food, and various other resources, are derived from wild species. The vast majority were domesticated several thousand years ago and since that time they have undergone a phase of rapid evolutionary change. Until the present century the selection was imposed 'unconsciously', in that our ancestors had no idea of what they were doing in terms of genetics. The selection, and 'breeding', of plants and animals was an art, rather than a science as we know it today. Scientific principles were only applied after the laws of heredity had been established, and when an understanding was gained on how to handle characters which show continuous variation.

Plants

The early phase of 'unconscious' selection in plants was so successful that there have been almost no new *staple* crops (those forming a main element in our diet) introduced in the world for the past 2000 years. Most of the major cereals, root crops, fruits and vegetables were established before that time, although their improvement has been going on continuously right up to the present day. An important exception is *Triticale*—a new allopolyploid cereal produced in the last three decades by hybridisation between wheat and rye.

The early phase of domestication was little more than natural selection—saving seeds from the biggest and best individuals, and planting out the ones that survived the ravages of various pests and diseases. Selection was done on phenotypes, of course, but over the centuries many small heritable variations have accumulated and given rise to some remarkable changes. The major part of the selection has been on characters which show continuous variation, but some progress was also due to major mutational events like the allopolyploidy which gave rise to our cultivated wheats (Chapter 16, Box 16.2). One example will serve to illustrate the range, and kind, of variation that can be obtained by selection from within a species: that of different cultivated forms derived from the wild cabbage (Fig. 21.1).

Fig. 21.1 Variation under domestication in the cabbage, *Brassica oleracea*. The wild cabbage is shown at the left (*a*). To the right are three cultivated forms of the species which have been modified by selection for different parts of the plant. The cauliflower (*b*) is a modified inflorescence comprising thousands of fleshy flowerbuds. Brussels sprouts (*c*) have been selected for axillary buds and the cabbage (*d*) for numerous overlapping leaves which surround the terminal bud. Three other varieties are not shown—broccoli, kale and kohlrabi. If they are allowed to flower all the different forms can be hybridised. (Photographs by courtesy of Mr P. T. Nelson, National Institute of Agricultural Botany.)

Animals

The first animal species to be taken into domestication by man was the dog (*Canis familiaris*). Remains dating from 10 000 BC have been found in Israel and Iraq. The dog is descended from the wolf and became our most widely distributed domestic animal. It was found to be useful for hunting, keeping flocks of animals, scavenging and also as a source of food. At the present time its main role is as man's favourite pet. There are now more than 500 distinct breeds of dog which have evolved over a period of time by selective breeding to suit different requirements. They all belong to the same species and can be interbred to give mongrels. The dog is probably the best example to show the extent of variation that can be produced by selection within a species of domestic animal—we only have to compare the St Bernard and the Chihuahua to appreciate the point. The way in which the anatomy has been modified by selection for different breed characters is illustrated in Fig. 21.2. Changes in the skull are particularly striking—no other species of animal shows such a large divergence in this character.

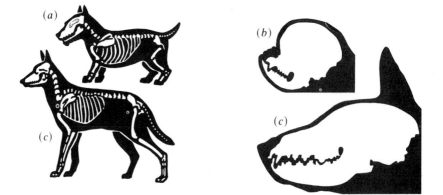

Fig. 21.2 Effect of selection on skeletal structure in the dog. Note the change in the leg bones following selection for miniaturising features in the Scottish terrier (*a*); and the alteration in the nasal bones of the short-faced pug (*b*) compared to the German shepherd (*c*).

Resistance to pesticides and antibiotics

Man is engaged in a constant struggle against other species which compete with him for food and other natural resources. An enormous part of the potential yield of our crops and animals products is lost each year to pests of one sort or another (various animals, insects, fungi, bacteria, weeds). To minimise these losses we combat the species concerned with a wide range of control measures. One of the commonest lines of attack is the use of chemical poisons, called 'pesticides'. The term **pesticide** is very general and includes *any substance used to control organisms which cause injury, disease or losses in growing crops, domestic animals or man*. We can also refer to particular kinds of pesticides, such as insecticides, herbicides, fungicides, nematocides, rodenticides, bacteriocides, etc.

An **antibiotic** is *a substance produced by a micro-organism which is able to inhibit the growth of other microbes, or to destroy them.* Antibiotics may be purified from the organisms which produce them naturally, or made by chemical synthesis. Many of these chemical controls are only partly effective, however, because in time it is found that their target organisms respond by evolving resistant strains which can overcome them.

Resistance, in this context, is *the ability of organisms to survive exposure to a normally lethal dose of a chemical poison.* Resistance develops because a certain proportion of individuals already possess resistance alleles to many different chemicals, even though they have never been exposed to them. This is a result of the random changes brought about in genes by mutation. The chemicals themselves do not cause mutations; they simply select out the resistant individuals which then survive to give progeny. The genetic composition of the pest population, and many other 'innocent' species, thus changes in response to selection imposed by man. There are today over 400 known cases of species that have developed resistance to insecticides, and somewhat lower figures for fungicides, bacteriocides and herbicides. The resistant forms have all become apparent in the last 35 years as a result of selection pressures imposed by the frequent application of high levels of pesticides. Some examples are detailed below.

Warfarin resistance in rats

Warfarin (3-α[acetonylbenzyl]-4-hydroxycoumarin) is an anticoagulant poison widely used as a **rodenticide** for the control of rats and mice. It was first introduced in 1950 and became a popular poison, against rats in particular, because of its low toxicity to farm animals. It acts by interfering with the way in which vitamin K is utilised in the normal process of blood clotting. When the bait is administered to sensitive rats the capillaries become much more fragile than normal, the blood fails to clot and the affected animals succumb to haemorrhages and slowly bleed to death—apparently without pain!

Resistant strains of the rat first appeared in Scotland in 1958. By 1972 resistant rats had appeared in twelve other areas of Britain. Breeding tests established that its genetic basis could be explained in terms of a single gene mutation in which resistance is conferred by a dominant allele. The gene is designated as **Rw** and its dominant and recessive alleles as **Rw^r** and **Rw^s** respectively. Evidently, the blood clotting process of the resistant strains is insensitive to Warfarin: clotting takes place in the normal time, but there is an enhanced requirement for vitamin K.

One resistant population that has been particularly well monitored is that in an area centred on Welshpool, in mid Wales, and extending over into the border of the English county of Shropshire (Fig. 21.3). Resistant rats were first noticed near Welshpool in 1959. Thereafter they were seen to spread out from the original site at a rate of three miles per year, which is the normal rate at which rats invade new territory. Their progress was dependent upon the continued use of the poison, and the constant selection pressure which was applied to the population, for several years. What happened here was that the environment of the rat, in terms of its diet, was suddenly and deliberately changed by the intervention of man's activities. This altered the selective value of the genes normally involved in the blood clotting process, and their action suddenly became lethal in the homozygous form (**Rw^{s}Rw^s**). A few variants carrying the dominant resistance allele were then favoured by selection, in the new environment, and the genetic composition of the population changed in response to the Warfarin baiting. In any wild population of rats, mutation will be constantly generating new alleles of the genes which mediate in the biochemistry of blood clotting—and some of them will confer resistance to Warfarin even though the population has never been exposed to the substance previously. Under normal circumstances these mutations would most likely be deleterious, as indeed they are in the case of rats in a natural environment, and selection would keep them down to mutation frequency.

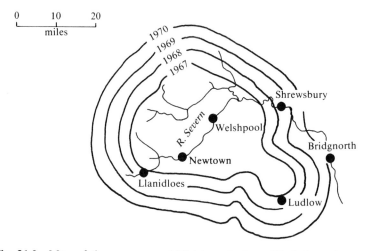

Fig. 21.3 Map of the area around Welshpool where Warfarin resistance in rats has been intensively studied. The map shows the spread of resistance over the three-year period 1967–1970. (Based on J. H. Greaves, and B. D. Rennison, 1973, *Mammal. Rev.* **3**, 27–29.)

The genetics of Warfarin resistance turns out to be more complicated, and more interesting, than it appeared at first sight. The complication arises from the fact that rats which are homozygous for the dominant resistance allele have a requirement for massive amounts of vitamin K, twenty times more than normal, which they have difficulty in meeting (vitamin K is synthesised by the gut bacteria). Normal homozygotes are poisoned by Warfarin, while the heterozygotes survive and require only a fraction more vitamin K than normal. Genotypes and phenotypes are as follows:

Rw^{s}Rw^s—normal rats, sensitive to Warfarin (recessive phenotype).

Rw^{s}Rw^r—heterozygotes, resistant to Warfarin and having a slightly increased requirement for vitamin K.

Rw^{r}Rw^r—resistant to Warfarin but having a twenty-fold increase in their vitamin K requirement, which greatly reduces their fitness compared to the heterozygotes.

The heterozygotes therefore have a selective advantage over both classes of homozygotes in the areas where warfarin is used, and this results in a **balanced**

polymorphism—as in sickle-cell anaemia (Chapter 20). Susceptibility to Warfarin in the recessive homozygotes (Rw^sRw^s) is balanced by the reduced fitness of the resistant homozygotes (Rw^rRw^r). Because of this the normal recessive allele can never be completely removed from the population.

Following the rapid spread of Warfarin resistance in the 1970s a new generation of related anticoagulant poisons was introduced. Difenacoum is one of the most widely used at the present time, but resistant populations are again emerging in some areas of Britain. Optimistic press reports declaring 'high noon for rats' are being superseded by more realistic headlines proclaiming 'super-rats overcome new poison'.

Fig. 21.4   The brown or common rat (*Rattus norvegicus*) is the most widely distributed of all rodent pests. It causes enormous damage by eating and spoiling both growing and stored food. In addition it damages the structure of buildings and harbours diseases that are transmitted to man and his domesticated animals. The rat is a prolific breeder and under ideal conditions one pair could theoretically produce 3000 individuals in one year. (Photograph by courtesy of M. R. Hadler, Sorex Ltd. See M. R. Hadler, 1984, Rodents and rodenticides, *Span* **27** (2), 74–76.)

DDT resistance in insects

This is the best known case of resistance to an **insecticide**. **DDT** (dichlorodiphenyltrichloroethane) was introduced about 50 years ago. It is cheap, easy to make and is a relatively stable substance:

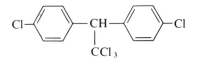

On a non-porous surface it will remain active for as long as eighteen months. In insects its main mode of action is on the nervous system; it causes loss of movement followed by paralysis and death. There are several mechanisms of resistance, determined by a number of different genes: (1) the nervous system itself may have a reduced toxic response to DDT; (2) the insect cuticle may be less permeable to DDT; (3) there may be an altered pattern of behaviour, so that the resistant strains avoid contact with the chemical; (4) there may be an increase in the insect's lipid content enabling the fat-soluble DDT to be separated from sensitive parts of the organism; and (5) some insects carry high levels of an enzyme that **detoxifies** DDT. In the housefly (*Musca domestica*) it is thought that many resistant strains contain high levels of the enzyme **DDT-dehydrochlorinase**, which breaks down the DDT into products which are less toxic (detoxification). Natural populations are pre-adapted in that there are always a few variants which carry high levels of the enzyme—even though they have never encountered DDT. Resistance is therefore the result of selection for these variants—they have higher levels of fitness and leave more progeny than the susceptible genotypes. Other important pests with resistance to DDT are the yellow fever mosquito (*Aedes aegypti*), and various species of malaria-carrying mosquitoes (*Anopheles* spp.).

Enormous quantities of DDT have been used over the years and much of it has accumulated in the fat deposits of animals for which it was never intended. It finds its way into food chains and then becomes concentrated in the bodies of predatory animals, including man. The use of DDT has now been abandoned in many parts of the world.

Insecticide resistance in peach–potato aphids

The peach–potato aphids (*Myzus persicae*, Fig. 21.5) have developed resistance to several different insecticides which are used against them in fields and greenhouses. They display what is known as **cross-resistance** to organophosphorus, carbamate and pyrethroid insecticides. Resistance which develops in response to any one of these substances will give protection against the others, and it appears to be based on greater activity of the enzyme **carboxylesterase** (**E4**). The form of the enzyme is the same in both resistant and susceptible

Fig. 21.5 The peach–potato aphid, *Myzus persicae*, viewed under the scanning electron microscope. (Photograph by courtesy of Rothamsted Experimental Station.)

strains, but the resistant ones produce it in much greater quantity. E4 degrades insecticides (and other substrates) by hydrolysing them. There are variants of the aphids which differ in the concentrations of the E4 enzyme that they carry, and these differences appear to be due to a series of *duplications* of the structural gene for the enzyme. The most resistant of the variants has 64 times as many copies of the gene as the susceptible strain, which has only one copy per haploid chromosome complement. Amplification of genes coding for E4 confers a selective advantage on the aphids in the presence of the insecticides. In Britain, susceptible strains have been almost entirely replaced by the resistant variant, which produces four times the normal quantity of E4. The most resistant variants with a 32 and 64-fold increase in E4 are mainly confined to greenhouses, where insecticides are used much more intensively.

Resistance to antibiotics in bacteria

Bacteria are important pathogenic organisms in man, and in domesticated animals. In human beings they are responsible for such serious diseases as tuberculosis, dysentery, urinary infections, typhoid fever and cholera, to mention but a few. Following the introduction of antibiotics in the 1940s and onwards, it was thought that many of these diseases would be quickly eradicated. This was not the case, however, and resistant strains of these disease-causing organisms quickly developed in response to the widespread use of antibiotics. Some of the most troublesome bacteria are *Staphylococcus* spp. which live in all our noses and are the cause of many serious infections— particularly in hospital surgical wards. Penicillin was highly effective against this bacterium when it was first brought into clinical use in 1940. Within ten years, though, the majority of staphylococcal infections in hospitals throughout the world were penicillin-resistant. The basis of the resistance is the enzyme **penicillinase** which cleaves the β-lactam ring of penicillin and renders it inactive. The beta-lactamase enzyme, and the gene which codes for it, have always been present in bacterial populations (penicillin is a natural substance after all). Strains which produced high levels of penicillinase were selected when penicillin became widely used. In the 1950s, new antibiotics were introduced, such as streptomycin, chloramphenicol, tetracycline, etc., but these in turn have all led to the emergence of resistance strains.

As described in Chapter 12, bacteria frequently carry **plasmids** as well as their main chromosome. These plasmids are small circular molecules of DNA which can be classified into several types. One class of plasmids, known as **R-plasmids**, carry genes for drug resistance, although not all resistance genes are necessarily carried in plasmids. These plasmids are self-replicating structures which are inherited when the bacterial cells multiply by binary fission. They can also be transferred from one strain to another, by conjugation, and this greatly facilitates the spread of resistance genes.

Infectious drug resistance is the term used to describe the way in which resistance genes can be transmitted by the plasmids during conjugation. In the late 1950s in Japan it was discovered that some strains of *Shigella*, the bacteria which cause dysentery, carried plasmids with several different resistance genes in them, i.e. they had **multiple drug resistance**. This multiple resistance had built up because the bacterial populations in hospitals had been exposed to several different drugs over a period of time. The problem is not confined to Japan, but is now common throughout the world. Because the genes concerned are in plasmids it means that they are all transferred together, during conjugation, from a multiple-resistant strain to a sensitive strain which lacks the plasmid. These R-plasmids can also be exchanged between different species of bacteria, and drug resistance can be passed, for example, from *Shigella* to the normal gut bacterium *E. coli*. One person can therefore pick up multiple drug resistance in *Shigella*, from another person, through *E. coli* as an intermediate. Multiple drug resistance built up in farm animals can also be transferred to bacteria carried by people working with them. At one time this posed a serious threat to public health, because farm animals were routinely fed on diets containing antibiotics in order to improve their growth rates. Legislation has now been passed to restrict the use of antibiotics in animal feedstuffs.

Summary

Man has deliberately interfered with populations of other species in order to feed himself and to gain protection from pests and disease-causing organisms. In the process of doing this he has imposed selection upon the species concerned. Wild plants and animals have been domesticated, over a period of several thousands of years, and changed by strong selection pressures out of all recognition to their original form. This response to man-made selection has provided convincing evidence of the power of selection to act upon the variation within a species and to lead to diversity and evolution. The use of pesticides and antibiotics is a recent phenomenon, and its effects have mainly taken place within the last 35 years. In an attempt to control or eradicate pests man has applied massive quantities of poisonous chemicals to the environment. The target organisms have responded by evolving mechanisms of resistance, and changing the genetic composition of their populations to adapt to the altered environments.

Further reading

Heiser, C. B. (1981), *Seed to Civilisation*, W. H. Freeman.

Questions

1 Housefly populations tend to exhibit an increasing resistance to insecticides such as DDT and this resistance is associated with the production by houseflies of the enzyme DDT-dehydrochlorinase. The resistance of mosquitoes to another insecticide, dieldrin, has developed at an even faster rate. Describe a possible mechanism by which the resistance of houseflies and mosquitoes has developed. Why may the resistance of mosquitoes have developed faster?   *(A.E.B. Biol., 1982)*

2 Explain how studies on drug resistance in bacteria contribute to our understanding of organic evolution.

3 (*a*) Give an account of the genetic basis of Warfarin resistance in rats.
   (*b*) How does the fact that resistance is determined by a dominant gene affect the spread of resistance.
   (*c*) Explain why the spread of resistance, in the Welshpool area for example, was dependent on the continued use of Warfarin.

22
Genetic engineering

Mendel's discoveries about the laws of heredity enabled us to begin manipulating the genetic material in a scientific way, and to change the genetic constitution of plants and animals to suit our needs. Crosses could be made in such a way that desired combinations of characters appeared among segregating progenies in predictable ratios. As more knowledge was gained on linkage, sex linkage, continuous variation, etc., our control over the processes of heredity improved. Developments in chromosome genetics (cytogenetics) enabled us to manipulate whole genomes and to make new kinds of hybrids and polyploids in crop plants. This work on the improvement of domesticated species goes on all the time, and has now reached a high level of technological sophistication. All such manipulation relies mostly upon natural processes—recombination, selection, hybridisation, and chromosome doubling (with colchicine). In the 1970s, however, a new development took place which completely revolutionised the way in which the genetic material could be handled. 'Restriction enzymes' were discovered. These enzymes were isolated from bacteria and were found to have the property of cutting DNA into discrete fragments. When used in conjunction with other enzymes which join up broken ends, it became possible to combine together pieces of DNA from totally unrelated species, and to make what is called 'recombinant DNA'. Recombinant DNA technology allows us to manipulate DNA in ways that are quite novel, and in some cases quite unnatural—such as placing human genes into bacterial chromosomes so that they function and make their gene products within the bacterial cells. This new facility has enormous potential for genetical research as well as for the experimental modification of organisms by 'genetic engineering'.

Genetic engineering

Genetic engineering is *the production of recombinant DNA molecules by the insertion of DNA into a virus, a bacterial plasmid or other vector system, and their incorporation into a host organism.* Some aspects of this new branch of genetics are discussed below.

Restriction enzymes

Virtually all species of bacteria synthesise enzymes which make double-stranded cuts in DNA. They apparently use these enzymes to protect themselves from foreign DNA which gets into their cells—mainly from bacterial viruses (Chapter 12). Because these enzymes restrict the range of host bacteria that a virus can invade they are known as **restriction enzymes,** or more precisely

restriction endonucleases. Restriction enzymes can be extracted and purified from cultures of bacteria and then used by the genetic engineer to make recombinant DNA. They are particularly useful due to the way in which they only cut the DNA at certain **sequence specific sites**. In the bacteria themselves some of the DNA bases in these same sites on the bacterial chromosome are modified by the addition of methyl ($-CH_3$) groups which protects them from degradation by the cell's own endonucleases.

Several hundred restriction enzymes have now been identified and isolated. They are each named after the bacterial strain from which they were derived, e.g.:

> *Eco* R1 is from *Escherichia coli*, strain RY13
> *Hin* dIII is from *Haemophilus influenzae*, strain Rd
> *Bam* H1 is from *Bacillus amyloliquefaciens*, strain H.

Most restriction enzymes recognise and cut DNA within specific nucleotide sequences, often 4 or 6 base pairs long, called **restriction sites**. These restriction sites are usually in the form of **palindromes**. That is to say they comprise nucleotide sequences that are symmetrical, about an axis, and read the same in opposite directions in the two strands of the DNA. *Eco* R1, for instance, makes double-stranded cuts in the way shown in Fig. 22.1. It produces a cut which is staggered, giving **sticky** or **cohesive** ends.

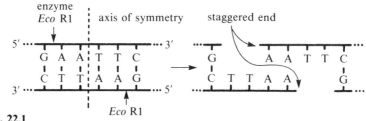

Fig. 22.1

Other restriction enzymes, such as *Hae* III, cut the DNA strands at positions directly opposite one another, giving **flush** or **blunt** ends.

Palindromes are found in the DNA of all species. They are located at particular sites which are randomly distributed along the chromosomes. In most species, as far as we know, they have no particular biological function. Their occurrence, however, is extremely useful to the genetic engineer, because they are the sites which are cut, or **cleaved**, by restriction enzymes. If samples of DNA from two different species are treated with the same restriction enzyme, say *Eco* R1 again, then they will both be cleaved at the *same* palindromic restriction sites and will have the *same* sticky ends to their fragments. When the two kinds of DNA fragments are mixed together, they will **anneal** (by complementary base pairing) and some new kinds of molecules will be produced. The rejoined ends can then be sealed by special joining enzymes, called **ligases**. These joining enzymes can also be used for sealing blunt-ended fragments.

It is this capability to fragment DNA into discrete pieces and then to join up different fragments together which forms the basis of **recombinant DNA technology** (Fig. 22.2). It means that we can manipulate DNA and move genes

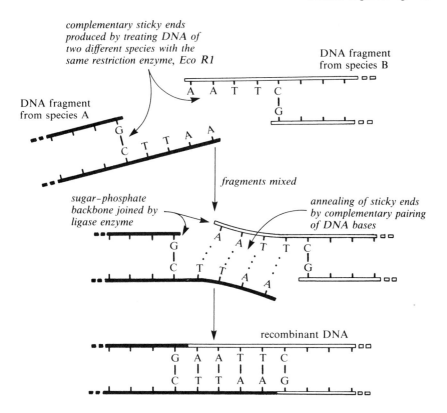

Fig. 22.2 Diagram showing how recombinant DNA is made.

around from one species to another, in order to engineer new kinds of organisms that will perform certain desired functions.

Recombinant DNA

The most widely used system for working with recombinant DNA at the present time is that involving the bacterium *E. coli* and its small DNA plasmids. Plasmids were discussed in Chapters 12 and 21. They are small circular molecules of double-stranded DNA which are found in many kinds of bacteria in addition to their main chromosomes, and which perform various functions. They replicate themselves independently of the main chromosome. In general they are not vital to the cell and can be used by the genetic engineer as vectors to carry fragments of DNA, or genes, from other species. Plasmids can be taken out of bacterial cells, modified in various ways as recombinant DNA molecules, and put back into bacteria which serve as hosts to allow them to multiply or express their genes.

A generalised scheme for making recombinant DNA plasmids in *E. coli* is illustrated in Fig. 22.3. The sequence of operations begins by obtaining plasmids from a culture of bacterial cells. The bacteria are broken open and the plasmids separated out by centrifugation. These plasmids usually contain single sites for a number of different restriction endonucleases. They also contain one or more genes for antibiotic resistance. The bacterial cells with the plasmid genes for antibiotic resistance will grow on a medium containing the antibiotic. The bacterial cells lacking the resistance genes will not grow. Cells which carry the plasmids can therefore be selected for by growth in a medium containing the antibiotic.

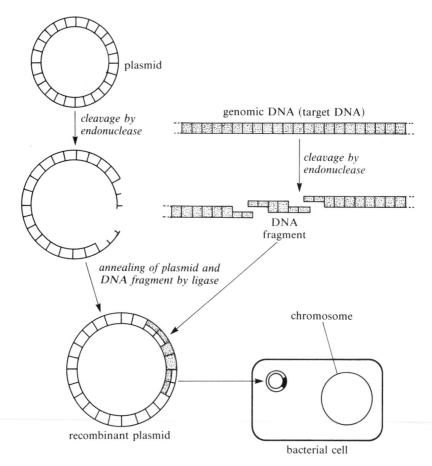

Fig. 22.3 Generalised scheme for making recombinant DNA plasmids in the bacterium *E. coli.* See text for explanation.

The plasmids are opened up at one of their restriction sites, by cleaving with the appropriate restriction enzyme, and then mixed with the desired DNA fragments which have been produced from the source which is required by the same restriction endonuclease. Because both plasmid and target DNA have identical sticky ends, they will form recombinant DNA molecules, in which some plasmids will come to contain a single fragment of the target DNA (Fig. 22.2). After their single-stranded sticky ends have annealed, the gaps in the sugar–phosphate backbone of the DNA are sealed by **DNA-ligase**. These recombinant plasmids are then mixed with host bacterial cells, some of which take in a single plasmid through their cell membrane. This transformation is aided by treatment with $CaCl_2$ which makes the host cells 'competent' for transformation. A number of cultures are then started from single cells, on media containing antibiotic, so that those which contain recombinant plasmids can be sorted out from those which carry non-recombinant plasmids. We will not go into the details of how this sorting out is done.

Using recombinant DNA plasmids

There are several ways in which recombinant DNA plasmids are being used in genetic engineering programmes:

Cloning   Once a particular restriction fragment of DNA, or a whole gene which is of some experimental or practical importance has been stably inserted into a bacterial plasmid it is said to have been **cloned**. To obtain a supply of the cloned fragment a large population of the bacteria are simply grown up in culture. The medium contains antibiotic to select against cells without any plasmids. Each bacterium in the culture therefore contains one or more of the self-replicating plasmids which carry the desired foreign DNA fragment. The cells are then broken open and their plasmids separated out by centrifugation.

Restriction enzymes are then used again to release the cloned fragments which can be utilised experimentally. Many eukaryotic genes have been cloned in this way in order to obtain them in large quantity for studies on their structure and DNA base sequence. It was as a result of this kind of work that the split genes described in Chapter 14 (Box 14.3) were discovered.

Synthesis of gene products   There are many valuable gene products which it is desirable to obtain in large quantity. Some of them are of commercial value, such as enzymes used in industry, while others have important medical applications, e.g. insulin, interferon, and somatostatin. These proteins can now be synthesised inside genetically-engineered bacterial cells and then extracted by various means. The human growth hormone somatostatin is an important example. It is a small protein of only 14 amino acids. The gene was synthesised in the laboratory and then inserted into a plasmid of *E. coli* in such a way that it could be 'switched on' and made to produce the human growth hormone inside the bacterial cells (Fig. 22.4 overleaf).

Transformation in eukaryotes

Eukaryotes are much more complex organisms than prokaryotes and it is correspondingly more difficult to modify their genetic material by recombinant DNA technology. The objectives are simple enough—insert genes into animals and plants that will make them more efficient producers or give them better heritable resistance against pests and diseases. Ultimately it may become possible to modify the genomes of man so that serious genetic disorders can be corrected by **gene therapy**.

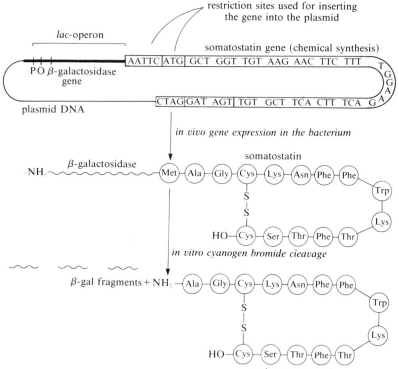

Fig. 22.4 Synthesis of the human growth hormone somatostatin using a synthetic gene inserted into a genetically-engineered plasmid of *E. coli*. The plasmid also contains part of the lactose operon of *E. coli* which will switch on the gene when lactose is fed to the cell culture (Chapter 17). After synthesis the cells are broken open to release the proteins and a simple chemical treatment then splits off the somatostatin from the β-galactosidase which is made with it. The somatostatin gene is represented as the *genomic sequence*—this is the sequence in the non-transcribed strand of the DNA which has the identical codons to those of the mRNA except that **T** is present in place of **U**. (Based on K. Itakura *et al.*, 1977, *Science* **198**, 1056–1063.)

In engineering new forms of plants and animals, there are three stages which must be accomplished in any successful programme:

1. Cloning of the desired gene and then multiplying it up in a bacterial culture to give an abundant supply; methods of obtaining genes are discussed below.
2. Placement of the new gene into the recipient nucleus; in the case of animals, this may be a fertilised egg which is manipulated *in vitro* and then re-implanted into a foster mother for development. With plants it is more difficult because the egg cells cannot be so easily manipulated in this way and there is the additional barrier of the rigid cellulose cell wall. One way around this problem is to work with **protoplasts**. These are cells that have had their cell walls removed by enzymes (Fig. 22.5). In some species, single protoplasts can be grown in culture and regenerated into whole plants (e.g. tobacco).

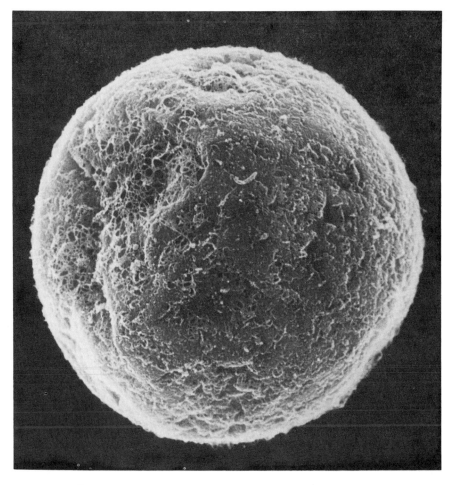

Fig. 22.5   An example of a plant protoplast as seen under the scanning electron microscope (×6000). This one is from a moss (*Physcomitrella patens*). The protoplast has been released by treatment of the 'leafy' tissue of the moss with an enzyme solution that strips off the cellulose walls from the cells. It has been cultured for four hours in a growth medium and is regenerating a cell wall. (Photograph kindly supplied by Dr Jeremy Burgess, John Innes Research Institute.)

3. Integration of the new genes into the recipient's own DNA, ideally at a particular site where the genes will be stably inserted and will function to make their protein products: this stage is important because it ensures that the transformed individual will have the new gene in all of its cells and will then transmit them to its progeny during the normal course of reproduction.

Successful transformations in animals (e.g. rabbits, mice, *Drosophila*) have now been achieved by micro-injecting genes directly into the nucleus of a fertilised egg cell (Fig. 22.6). Other methods which are being experimented with are simply mixing DNA fragments with eggs to get passive uptake of the genes, or inserting genes into viruses or transposable elements which act as vectors to take the genes into the host cells. Mixing DNA fragments with egg cells is the simplest way, but it is also the least efficient, since only a small proportion of eggs will actually take up the DNA—and fertilised animal eggs are not all that easy to come by anyway. By micro-injection the genes can be placed directly into the nucleus. The use of certain viruses as vectors is an attractive idea, because they naturally insert their DNA into the host chromosomes, taking the new genes with it. Once inserted, the viral DNA will be replicated along with the chromosomal DNA. The hazard with this approach is that the viruses themselves may do harm to the recipient host cells.

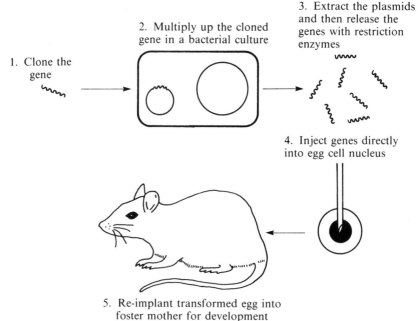

1. Clone the gene

2. Multiply up the cloned gene in a bacterial culture

3. Extract the plasmids and then release the genes with restriction enzymes

4. Inject genes directly into egg cell nucleus

5. Re-implant transformed egg into foster mother for development

Fig. 22.6 A method for the introduction of new genes into animals.

An example of a successful attempt at genetically engineering a new strain of an animal is that of the 'Mighty Mouse'. A gene known as **GH**, which produces a growth hormone in the rat, was cloned and joined up with a promoter region of another gene which was switched on by the presence of certain metals in the diet. The new gene was then micro-injected into fertilised mouse eggs which were subsequently re-implanted into foster mother mice. Transformed mice were found among the progeny. The new genes had become integrated into their chromosomal DNA and were actively producing rat growth hormone in various tissues of the body. Some of the mice grew to twice their normal size and some were also able to transmit their new genes to their offspring when mated with normal strains of mice. This kind of experiment holds out great promise for the future genetic modification of farm animals by recombinant DNA techniques.

Plants are much more difficult to manipulate, as explained above, and one of the avenues which is being explored at the present time is that of transformation using a **natural vector**. The potato plant, for example, is attacked by a bacterium called *Agrobacterium rhizogenes*. This is a soil-borne pathogen which invades the roots of the potato at sites of wounding and causes the formation of tumours. It gives rise to a condition known as 'hairy root disease'. The bacterium is a natural vector because it carries a plasmid, as well as its main chromosome, and when infection takes place part of this plasmid (called the tumour-inducing DNA or T-DNA) becomes integrated into the host cell chromosomes. This T-DNA, in fact, is the only part of the pathogen that is transferred into the host. Once integrated into the host chromosome, the T-DNA transforms the infected cells and exploits them to produce its own gene products—including tumour-specific compounds such as **opines** and **nopalines**. These tumour-specific products are then secreted from the tumour cells and used as a source of nutrition by the *Agrobacterium* cells living in the soil surrounding the tumours.

Fig. 22.7   A normal potato plant (right) and a plant transformed by the T-DNA from a plasmid of *Agrobacterium rhizogenes* (left). In the transformed plant the T-DNA is integrated into the chromosomes of all the cells. Note the glossy crinkled leaves and the altered morphology resulting from expression of the T-DNA genes. The transformed plant was obtained by regeneration from a piece of hairy root tissue of an experimentally infected plant. This plant also has an extensively branched root system. (Photograph kindly supplied by Dr Gert Ooms, Biochemistry Department, Rothamsted.)

Genetic engineers are obviously keen to make use of this natural system of transformation as a means of placing desired genes into crop plant genomes. Another *Agrobacterium* species, *A. tumefaciens* which induces crown gall development in infected plants, is also being considered for this purpose. Both species of *Agrobacterium* are able to infect a wide range of dicotyledonous plants. The idea that is being developed is to modify the T-DNA of the vector plasmids in such a way that (1) the tumour-inducing genes are removed, (2) the T-DNA will retain its capacity to integrate into plant chromosomes and (3) the T-DNA will carry new genes to modify and improve the crop plants concerned. It will be necessary, of course, to transform plants in such a way that the T-DNA of the engineered plasmids is stably integrated into the chromosomes in all of the cells, and can then be inherited as a permanent part of the genome. For this to be attained, whole plants must be regenerated from transformed tissues.

Success in this area has already been achieved at the experimental level. Tim Hall has used an engineered form of the tumour-inducing plasmid of *A. tumefaciens*, containing a storage protein gene from bean, to transform tobacco plants. Tobacco plants regenerated from transformed tissues express the bean storage protein genes in their seeds. (Tobacco is widely used in these kinds of experiments because of the ease with which whole plants can be regenerated from single protoplasts.) Another major advance in this programme has recently been made by Gert Ooms. Ooms has succeeded in regenerating whole potato plants from hairy root tissues which carry the T-DNA of the natural *A. rhizogenes* plasmid. The T-DNA is carried in all of the cells of the transformed plants which have a modified phenotype due to expression of the T-DNA genes (Fig. 22.7). Neither of these experiments has produced agri-culturally-useful plants, but they have revealed the means by which this might be achieved.

Obtaining genes

Eukaryotic genomes are extremely large and contain many millions of nucleo-tide base pairs. How can we find and isolate a single gene from so much DNA?

One way is not to try. In the case of a small protein, such as somatostatin or insulin, the amino acid sequence of the polypeptide is known and this tells us the genetic code of the gene concerned (Chapter 14). The gene sequence is then made by chemical synthesis. Automated machines are available which can be programmed to assemble DNA nucleotides to order, and can be used for the manufacture of small genes. Once a few copies have been made they can then be cloned in the usual way. In time it may be possible to synthesise any gene that we require by this means.

Another method of obtaining individual genes is by identifying and isolating their mRNAs. In certain cells of an organism only one gene, or a few genes, may actually be working and producing mRNA. This is the case in the reticulocytes of the bone marrow which carry mainly the mRNA for haemo-globin (p. 235). In such cases, the mRNA can be extracted and then used as a template on which to make a complementary DNA (cDNA) by reverse tran-scription. This is done with the aid of an enzyme called **reverse transcriptase** which will produce a single-stranded DNA from mRNA. The cDNA is then made double-stranded with the enzyme DNA polymerase. A cDNA copy of the gene is therefore obtained indirectly by reverse transcription from its own messenger RNA. There are also ways of sorting out a single desired mRNA from a whole mixture of mRNA transcripts, and it is now possible to obtain cDNA copies of almost any gene even when its mRNA is not abundant within the cells. In eukaryotes, as explained in Box 14.3 (p. 203), these messengers will have undergone a 'processing stage' before leaving the nucleus and certain sequences in them, which correspond to the non-coding introns of the genomic DNA in the chromosomes, will have been spliced out. Furthermore, the mRNA transcripts do not contain the sequences at the beginning of the gene which are necessary for transcription, because these sequences are not themselves transcribed into mRNA (i.e. sites for ribosome binding and attachment of RNA polymerase). The cDNA copy of a eukaryotic gene is therefore different from that of its genomic sequence.

If the genomic sequence itself is required, it can be isolated from a mass of DNA fragments by using a radioactive cDNA or mRNA **probe** which will hybridise to the corresponding genomic sequence by complementary base pairing.

Summary

The discovery of restriction enzymes in the 1970s opened up a completely new field of genetics. The restriction endonucleases are isolated from bacteria which use them in defence against viruses which invade their cells. They cause double-stranded cuts in DNA and when used in conjunction with ligases, which are joining enzymes, they enable us to make completely novel forms of recombinant DNA. Genes can be removed from one species and inserted into the DNA of another one so that new kinds of genetically-engineered DNA molecules, and hence new kinds of organisms, can be made. Bacterial plasmids have proved particularly useful as vectors for cloning pieces of DNA and for carrying foreign genes inside bacteria so that their cells may be used as factories for the manufacture of desirable gene products. The main objective now is to use recombinant DNA technology to transform higher plants and animals and to modify them to produce better crops and more productive farm animals. It is also hoped that these techniques can be used in gene therapy in order to correct certain genetic defects in man.

Question

1 Explain the phenomena of cloning and genetic engineering and discuss the significance of these developments.   *(L. Biol., Special Paper, 1983)*

Glossary of terms

Acrocentric
A chromosome with the centromere located near to one end, making one chromosome arm much longer than the other.

Adaptation
The structural or functional features that have evolved in an organism and which enable it to cope better with its environment.

Adaptive radiation
The evolution of a group of organisms along several different lines involving adaptation to a variety of environments.

Allele (allelomorph)
A particular form of a gene occupying the same locus (=place) on a chromosome as alternative alleles of the same gene.

Allele frequency
See *gene frequency*.

Allopatric
Of populations or of species that inhabit separate geographic regions (in contrast to *sympatric*).

Allopolyploidy
Polyploidy due to multiple sets of chromosomes from more than one species.

Alternation of generations
The regular alternation of haploid and diploid phases in the life cycle of sexually reproducing plants.

Aneuploidy
Change in chromosome number involving only part of a chromosome set.

Antibody
A protein (immunoglobulin) which is usually found in serum and whose presence can be demonstrated by its specific reaction with an antigen.

Anticodon
The triplet of nucleotides in a transfer RNA molecule that associates by complementary base pairing with a specific triplet (codon) in the messenger RNA during protein synthesis.

Apomixis
Seed production without fertilisation.

Autopolyploidy   Polyploidy due to the presence of more than two sets of chromosomes from within a single species.

Autosomes   The chromosomes of a complement other than those involved in sex determination.

Auxotroph   A mutant strain of an organism that is unable to synthesise a given organic molecule, and can only grow when the required substance is supplied in the food.

Backcross   A testcross in which an individual is crossed with one of its parents.

Bacteriophage (phage)   A virus which infects a bacterial cell and reproduces inside it.

Balanced polymorphism   A stable genetic polymorphism that is maintained by natural selection.

Balbiani ring   See *puff* (*chromosome puff*)

Barr body   A condensed inactivated X chromosome which is visible in the interphase nucleus.

Bivalent   Two homologous chromosomes that are paired together during the first division of meiosis.

Blending inheritance   An early theory of heredity (now discredited) in which it was thought that the characters of an individual were formed as a result of the blending of essences from its parents.

Breeding system   The method of breeding, by which we mean whether an organism is inbreeding or outbreeding.

Carrier   An individual with a mutant allele which is not expressed in the phenotype because of the presence of a dominant allele.

Cell cycle (mitotic cycle)   The life cycle of an individual cell.

Centromere   A region of a chromosome that becomes associated with the spindle fibres during mitosis and meiosis.

Character   Any observable phenotypic feature of a developing or fully-developed individual.

Chiasma (*plural* chiasmata)   The cross-shaped arrangement of the chromatids which is formed at their point of exchange during crossing over. Chiasmata are visible from diplotene until the beginning of anaphase I.

Chromatid   Each of the longitudinal subunits of a duplicated chromosome that become visible during mitosis and meiosis.

Chromatin   The stainable material in the nuclei of cells which is composed of DNA and proteins.

Chromomere   One of many bead-like structures visible along a chromosome during prophase of meiosis and mitosis.

Chromosome   A thread-like structure found within the nuclei of cells and containing a linear sequence of genes. Eukaryotes have several chromosomes composed of a complex of DNA and protein, whereas prokaryotes have a single chromosome of naked DNA.

Clone   A population of cells or organisms derived by mitosis from a single cell or common ancestor.

Cloning
(molecular cloning) The production of a number of identical DNA molecules by replication of a single DNA fragment in a suitable host system (such as a bacterial plasmid).

Co-dominance   A condition in which both alleles of a gene are expressed in a heterozygote which then exhibits the relevant characteristics of both parents.

Codon   A triplet of nucleotide bases in the DNA that codes for a specific amino acid or for polypeptide chain termination during protein synthesis.

Conjugation   The unidirectional transfer of genetic information between donor and recipient bacterial cells involving direct cellular contact.

Coupling   Two linked heterozygous gene pairs in the arrangement $\dfrac{AB}{ab}$.

Crossing over   A process of exchange between homologous chromosomes which may give rise to new combinations of characters. It takes place by the breaking and rejoining of chromatids and leads to the formation of chiasmata.

Cytokinesis   Division of the cytoplasmic component of the cell, and the separation of daughter nuclei, to separate cells.

Cytoplasm   The protoplasmic contents of a cell, excluding the nucleus.

Cytoplasmic
inheritance   The inheritance of characters determined by genes that are not located in the chromosomes.

Deletion	A mutation involving the loss of genetic material. The size of a deletion may vary from a single nucleotide to a chromosome segment.
Deme	See *Mendelian population.*
Dihybrid cross	Cross between homozygous parents that differ in two pairs of alleles (**AABB × aabb**). Likewise— trihybrid and polyhybrid.
Dioecious	Of plants with staminate and pistillate flowers on separate individuals.
Diploid	Cells, phases of the life cycle, and organisms which are characterised by having two sets of chromosomes.
DNA	Deoxyribonucleic acid. A polynucleotide in which the sugar residue is deoxyribose.
Dominant	An allele which masks the expression of another allele of the same gene. Also, the character produced by a dominant gene.
Double recessive	An individual recessive for two pairs of alleles (**aabb**).
Down's syndrome	A syndrome characterised by behavioural, physiological and mental defects due to chromosome mutation. The commonest cause is trisomy for chromosome 21.
Duplication	A mutation which results in doubling up of part of the genetic material the size of which may vary from a single nucleotide to a chromosome segment.
Endomitosis	Duplication of the chromosomes without cell or nuclear division.
Endopolyploidy	Increase in the number of chromosome sets that occurs when replication takes place without cell division.
Enzyme	A protein which speeds up, or controls, chemical reactions in living organisms without being used up during the reaction.
Episome	A genetic element in bacteria that can replicate free in the cytoplasm or which may insert into the main chromosome and then replicate along with the chromosome.
Epistasis	Where one gene hides the expression of another gene.

Euchromatin   A chromosome region that has normal staining properties and undergoes the normal cycle of coiling.

Eukaryote   A cell or organism that contains a true nucleus enclosed within a membrane.

Euploidy   Variation in the number of whole sets of chromosomes.

Evolution   The gradual change in the genetic composition of a population over a number of generations.

Fertilisation   The fusion of two gametes of opposite sex to form a zygote.

Fertility factor (F-factor)   An episome that determines whether a bacterium will act as a donor (F^+) or recipient (F^-) cell during conjugation.

Fitness (Darwinian fitness, relative fitness)   The relative reproductive contribution of a genotype, compared with another, to the following generations.

Founder effect   Change in gene frequencies that can occur when a new population is founded by a small number of individuals that are not representative of the genetic composition of the population from which they derive.

Frameshift mutation   A mutation resulting from the addition or deletion or one or more nucleotides, other than in multiples of three, that causes the gene to be misread.

Gamete   A mature reproductive cell which is capable of fusing with a similar cell of opposite sex to give a zygote.

Gametophyte   The haploid sexual generation which produces the gametes in plants with a haplo-diplontic life cycle.

Gene   A unit of heredity. A gene occupies a specific site in a chromosome and comprises a segment of DNA double helix about 1000 base pairs long that codes for an RNA or polypeptide product.

Gene flow   The transfer of genes between populations by the movement of gametes, individuals, or groups of individuals, from one population to another.

Gene frequency (allele frequency)   The relative proportions of the alleles of a gene present in a population.

Gene pool	The sum total of the genes within the reproductive cells of all of the members of a Mendelian population.
Genetic code	The way in which the genetic information is encoded in DNA.
Genetic death	The elimination of segregating individuals which are homozygous for deleterious recessives, or which carry deleterious dominant alleles.
Genetic drift	The random fluctuation in gene frequencies due to sampling error.
Genetic engineering	The production of recombinant DNA molecules by the insertion of DNA into a virus, a bacterial plasmid, or other vector system and incorporation of the recombinant DNA into a host organism.
Genetic load	The extent to which the burden of deleterious alleles causes a population to depart from its optimum fitness.
Genetics	The science of heredity. Geneticists study the transmission, the structure and the action of the material in the cell which is responsible for heredity.
Genome	The genes in the basic set of chromosomes.
Genotype	Genetic constitution of an individual.
Genotype frequency	The relative proportions of the various genotypes, from the alleles of one gene, present in a population.
Germ line	The lineage of cells from which the gametes are produced in animals.
Gynandromorph	A sexual mosaic which is typically male in certain parts of the body and female in others.
Haploid	Cells or organisms, or phases in the life cycle of organisms, with a single chromosome set.
Hardy–Weinberg Law (or Principle)	A principle which states that in a large randomly mating population there is a fixed relationship between gene and genotype frequencies and, in the absence of selection, migration and mutation, these frequencies remain constant from generation to generation.
Hemizygous	Pertaining to genes present only once in the genotype.

Heredity   The process that brings about the similarity between parents and their offspring.

Hermaphrodite   An organism which produces both male and female gametes.

Heterochromatin   Chromosome regions that are densely staining.

Heterogametic   The sex which produces two kinds of gametes with respect to the sex chromosomes.

Heterozygote   An individual carrying unlike alleles of a gene (**Aa**).
(*adj.* **heterozygous**)

Homogametic   The sex which produces only one kind of gamete with respect to the sex chromosomes.

Homologous   Chromosomes that are identical in their shape, size
  chromosomes   and the content and distribution of their genes (although different alleles may be present at the loci).

Homozygote   An individual carrying two identical alleles of a
(*adj.* **homozygous**)   gene (**AA** or **aa**).

Hybrid   An individual resulting from a cross between two genetically unlike parents.

Inbreeding   Any deviation from random mating, i.e. towards mating between relatives.

Incomplete dominance   A condition of a heterozygote in which the
  (semi-dominance,   phenotype is intermediate between the two
  partial dominance)   homozygous parental forms.

Independent   Separation of two or more unlike pairs of alleles
  segregation   independently of one another into the gametes as a result of meiosis.

Inversion   The reversal of the gene order which may result when two breaks occur in the same chromosome.

Karyogamy   The fusion of nuclei.

Karyotype   The chromosome complement of an individual defined by the number, the form and the size of the chromosomes at metaphase of mitosis.

Klinefelter's   A syndrome in man resulting from the presence
  syndrome   of an extra X chromosome in the male (XXY, XXXY).

Ligase   A joining enzyme which closes single-strand breaks in DNA.

Linkage	The association of certain genes in their inheritance.
Linkage group	A group of gene loci which are in the same chromosome.
Linkage map	A chromosome map containing in linear order the genes belonging to the linkage group associated with that particular chromosome.
Locus	The site in the chromosome where a gene is located.
Maternal inheritance	A form of inheritance controlled by genes outside the nucleus (as in the chloroplasts), in which the progeny have the genotype and phenotype of the female parent.
Meiosis	The reduction division of the nucleus in which the zygotic number of chromosomes is reduced to the gametic number
Mendelian population (deme)	A local community of a sexually reproducing species in which there is random mating, so that each member has an equal chance of mating with any other member of the population.
Messenger RNA (mRNA)	A single-stranded RNA molecule that is formed by transcription and which carries the information encoded in the gene to the sites of protein synthesis on the ribosome.
Metacentric	A chromosome having its centromere in the middle, and therefore having arms of equal length.
Migration	See *gene flow.*
Missense mutation	A base pair substitution in a gene which results in one amino acid being changed for another one at a particular place in a polypeptide.
Mitosis	Division of the nucleus into two daughter nuclei that are genetically identical to one another and to their parent nucleus.
Mitotic cycle	See *cell cycle.*
Monoecious	Of plants with both staminate and pistillate flowers on the same individual.
Monohybrid	Progeny of a cross between homozygous parents that differ in one pair of alleles ($AA \times aa$).
Multiple alleles	The alleles of a gene which has more than two.
Multiple genes	See *polygenes.*

Mutagen	An agent (radiation or chemical substance) that is capable of increasing the mutation rate.
Mutant	An organism, or a cell, carrying a mutation.
Mutation	A sudden heritable change in the genetic material.
Mutation frequency	The frequency with which a mutation is found in a sample of cells or individuals.
Natural selection	A process that determines gene frequencies in populations through unequal rates of reproduction of different genotypes.
Non-disjunction	A failure of chromosome separation in which a pair of chromosomes, or chromatids, pass to one pole of a cell instead of to opposite poles.
Nonsense mutation	Any mutation (substitution, addition or deletion) that changes an amino acid specifying codon into a chain terminating codon.
Nucleolus	A discrete structure in the nucleus of eukaryotes that is associated with the chromosomal site (nucleolus organiser region) of the genes coding for ribosomal RNA.
Nucleolus organiser	A region, or regions, in one or more chromosomes of the set which is associated with the nucleolus and which carries the rRNA genes.
Nucleoprotein	A molecule made up of a complex of nucleic acid and protein. The principal component of the chromosome fibre.
Nucleosome	A small spherical body which is the basic unit of eukaryotic chromosome structure. It comprises eight histone molecules encircled by two coils of DNA.
Nucleotide	A molecule composed of a nitrogen base, a sugar and a phosphate group—the basic building block of nucleic acids.
Nucleus	The membrane-bounded organelle of eukaryotes that contains the chromosomes.
Operon	A group of coordinately expressed and adjacent structural genes, together with their operator and promoter sites.
Outbreeding	Random mating.

Palindrome	A nucleotide sequence that is symmetrical about an axis of symmetry and reads the same in opposite directions in the two strands of DNA.
Parthenogenesis	Production of an embryo from an unfertilised egg.
Pedigree	A diagram of a family tree over a number of generations showing how the ancestors and descendants are related to one another.
Phage	See *bacteriophage*.
Phenotype	The appearance and function of an organism as a result of its genotype and its environment.
Plaque	A clear area in a bacterial lawn which results from the lysis of adjacent bacterial cells by successive cycles of bacteriophage reproduction.
Plasmid	A circular DNA molecule that replicates independently of the main chromosome within a bacterial cell.
Plastid	A self-replicating organelle in plants that can differentiate into a chloroplast.
Pleiotropy	The multiple phenotypic effects of a single gene.
Point mutation	A mutation which maps to a specific locus.
Polygenes (multiple genes)	Multiple genes (three or more) with individual small cumulative effects which control characters showing continuous variation.
Polygenic inheritance	Inheritance of characters whose expression is controlled by several genes with individual slight effects upon the phenotype.
Polymerase	An enzyme which catalyses the assembly of nucleotides into RNA or DNA on a DNA template during transcription.
Polymorphism	The presence of two or more distinct forms of a species (or of a gene) found in the same locality at the same time, and in such frequencies ($>1\%$) that the presence of the rarest form cannot be explained by recurrent mutation.
Polypeptide	A chain of linked amino acids.
Polyploid	A cell, tissue or organism with three or more complete sets of chromosomes.
Polytene chromosomes	Giant chromosomes which arise by endomitosis, in which the chromosomes replicate repeatedly without any separation of their chromatids.
Population	A local community of a sexually-reproducing species in which the individuals share a common gene pool.

Primary constriction   See *centromere*.

Prokaryote   An organism whose cell does not contain a true nucleus bound by a membrane.

Promoter   A DNA sequence at the beginning of a transcription unit that acts as a binding site for RNA polymerase.

Prophage   A temperate bacteriophage inserted as part of a host bacterial chromosome.

Protein   An organic compound composed of amino acids which are linked together by peptide bonds to form long chains, or polypeptides.

Protoplast   A plant cell without its cellulose cell wall.

Prototroph   A strain of organism that will proliferate on a minimal medium, in contrast to an *auxotroph*.

Puff (chromosome puff)   An expanded region at a site in a polytene chromosome where transcription is taking place.

Pure line   Descendants, by self-fertilisation, of a single individual which is homozygous at all of its gene loci.

Reading frame   The way in which the genetic code in mRNA is read as triplets by the ribosome.

Recessive   An allele whose effect is masked by the presence of a dominant allele of the same gene. Also, the character produced by a recessive gene in the homozygous state (**aa**).

Recombinant   An individual with a new set of characters arising by recombination.

Recombinant DNA   Novel DNA sequences spliced from DNA derived from more than one source.

Recombination   The process by which new combinations of parental characters arise.

Regulatory gene   A gene that regulates the activity of another gene.

Replication   The synthesis of DNA.

Repressor   A protein that binds to an operator site and prevents transcription of an operon.

Repulsion   Two linked heterozygous gene pairs in the arrangement $\dfrac{\mathbf{Ab}}{\mathbf{aB}}$.

Restriction enzyme   A nuclease that makes double-stranded cuts in DNA at specific palindromic sites.

Restriction site	Specific nucleotide sequence in DNA that is recognised and cut by a restriction enzyme.
Ribonucleic acid (RNA)	A single-stranded nucleic acid which is similar to DNA but having the base uracil (**U**) in place of thymine (**T**), and a ribose sugar instead of deoxyribose.
Ribosomal RNA (rRNA)	A class of RNA molecules that are present in ribosomes.
Ribosome	A complex structure composed of RNA and proteins which catalyses the translation of mRNA.
Samesense mutation	A base substitution in a DNA triplet which does not alter the amino acid sequence of a polypeptide.
Secondary constriction	See *nucleolus organiser.*
Segregation	Separation of pairs of unlike alleles (**A** and **a**) into different cells as a result of meiosis.
Sex-chromosomes	The chromosomes in the complement which carry the sex-determining genes.
Sex-limited	Genetically controlled characters that are expressed in only one sex.
Sex linkage	The location of a gene in a sex chromosome.
Somatic cell	A 'body cell' that is not destined to become a gamete.
Speciation	The process by which new species arise.
Species	Groups of interbreeding natural populations that share common morphological characteristics and are reproductively isolated from other such groups.
Spindle	A barrel-shaped structure made of microtubules that is formed in a cell during division of the nucleus, and which serves to align and move the chromosomes at metaphase and anaphase.
Sporophyte	The spore-producing generation in a haplo-diplontic life cycle.
Structural gene	A gene which codes for a functional protein.
Supergene	A group of linked genes that are inherited as a unit.
Sympatric	Of populations or of species that inhabit, or partly inhabit, the same geographic region (in contrast to *allopatric*).

Synapsis   Close pairing of homologous chromosomes at meiosis.

Telocentric   A chromosome with its centromere at one end.

Template   A molecular mould for the synthesis of a complementary molecule.

Testcross   A cross in which a heterozygote is crossed with a corresponding recessive homozygote (**Aa** × **aa**).

Totipotency   The capacity of differentiated somatic cells to retain their potential to produce an entire organism.

Transcription   The process in which messenger RNA is synthesised from a DNA template.

Transduction   The transfer of genetic material from one bacterium to another by a virus.

Transfer RNA (tRNA)   Small RNA molecules that carry specific amino acids to the ribosome during protein synthesis.

Transformation   The exchange of genetic information brought about by the uptake of naked DNA by a recipient cell.

Translation   The process by which the transcribed information carried in the base sequence of messenger RNA is used to produce a sequence of amino acids in a polypeptide chain.

Translocation   The transfer of a segment of one chromosome to another, non-homologous one.

Trisomic   A cell, tissue or organism having one of its chromosomes present three times.

Turner's syndrome   A sex-chromosome abnormality in man in which there is only a single X chromosome. Affected individuals are phenotypically female and usually have underdeveloped gonads.

Univalent   An unpaired chromosome at meiosis.

Virus   An infectious subcellular particle that requires a living host for its reproductive cycle.

Wild-type   The prevailing genotype or phenotype that is found in nature, or in a standard experimental strain of a given organism.

Zygote   A cell formed by the fusion of two gametes.

Index